Micro-optics and Lithography

Maria Kufner, Stefan Kufner

Micro-optics and Lithography

VUBPRESS

Cover photo: Lucia Feilner
Cover design: Danny Somers
Book design: B. Bardyn

© 1997 VUBPRESS - Brussels
ISBN 90 5487 167 9
NUGI 812
D / 1997 / 1885 / 004

VUBPRESS – VUB University Press, Pleinlaan 2 – B 1050 Brussels – Belgium
Fax + 32 2 6292694 – e-mail: kvschare@vnet3.vub.ac.be

Table of Contents

Table of Contents

Table of Contents

To Karl Oberberger

Introduction

Preface

Micro-optics is a special branch in optics which is strongly related to micro-electronics and also to the architectural aspects of computer science. Between the disciplines there is an exchange of concepts and technologies. Nowadays computers are designed and fabricated by electronics and computer science people only. Optics play a minor role in this field (eg. CD ROM). Nevertheless the speed of light and the enormous bandwidth make optics very attractive for information processing. Major developments have been achieved in the last three decades in optical technology, eg. laser technology, optical fiber technology, semiconductor optical devices, etc.

In the 60's optics already started an initiative for information processing, which had its beginning with the invention of the laser. But this was not a real challenge for electronics. One of the most important reasons for this is the lack of an optical transistor. For electronics this invention was made in the beginning and integrated circuits are like a bottom up design growing with the years of experience. Optical information processing, in contrast, started with concepts and architectures, while the basic switching element was still missing. Very early it was clear that an optical transistor or more general an optical switch will be a hybrid element having electronical and optical functions. Nowadays many such hybrid elements are known from literature and optics is to become a promising new technology for high speed information processing. Electronics on the other hand is also becoming faster, cheaper and with less power consumption per device and does not stop its own development as predicted especially for minimum feature sizes in lithography. For that reason many optical engineers are concentrating on a special part of information processing were electronics has a very weak point: interconnects.

Electronical architectures are laid out on boards. Wires may not cross in the plane and must have a certain distance between each other to prevent electromagnetic interference. Here optics offers a totally different approach which is known as parallel optical concept. Light can propagate through space without interferring. Thus the whole space in its three dimensions can be used as an interconnection system. From that a new architectural design arises. Information can be transmitted in parallel in two-dimensional arrays. The arrays can be interconnected between electronic and optoelectronic logic and switching devices through free space. This can help electronics to overcome the bottleneck of data flow and helps at the same time optics to become part of the computer world (eg. chip to chip or board to board interconnects). This requires also passive optical components like lenses, prisms, beam splitters and mirrors. The new challenge for optics arises in the miniaturization of these components to match optics to micro-electronics and to reduce space and therefore length of information channels which is equivalent to increasing the speed of data flow. Lithography is well known for integrated circuit manufacturing but it offers also a new potential for the fabrication of miniaturized optical components if modified techniques are applied.

Scope

This book provides an introduction into techniques that use lithography for the fabrication and integration of miniaturized passive optical components. It is directed to engineers and scientists who require practical and technical information on micro-optical engineering. It also serves as a textbook for a course in electrical engineering given at the Vrije Universiteit Brussel for the 5th year of students education.

Included are:

- diffractive optics
- waveguide optics
- refractive optics

A main aspect is the propagation of light in so-called free space optics. Not included are fabrication techniques for semiconductor optical devices and all kind of active components.

Approach and organization

The book starts with general aspects of miniaturizing optical elements. The transition from macro-optics to micro-optics and the included scaling behaviour is discussed in the first chapter.

Chapter 2 gives an introduction into the field of lithography which is the basic of all subsequent technologies. General aspects of polymers and their characteristics are discussed as far as they explain the physical side of the processes.

In the following chapters technologies for the fabrication of different types of micro-optical elements are discussed. For each technology there are examples of systems where this technology has been applied.

Chapter 3 gives a brief overview of diffractive optics. It has adapted the lithographic design modules from electronics like transferring very small feature sizes to a substrate by etching techniques. Here especially free space light propagation is emphasized which is a different approach to the propagation of information guided in wires.

The guided wave concept is explained in the following chapter 4 and the way in which it has borrowed some architectural concepts from electronics. Wave propagation in guiding layers is very similar to electronic concepts and suffers from almost the same restrictions to the layout.

Techniques for the miniaturization of refractive optics are subject of the next chapters. Chapter 5, 6, and 7 present different lens fabrication techniques: Photolythic methods, ion diffusion and melting techniques which are in use for the fabrication of various refractive lens types.

Chapter 8, 9, and 10 introduce technologies related to deep etch lithography which is a less mature field of research. Excimer laser light, synchrotron light and ion beams are used to create structures of several hundreds of micron thickness which is in contrast to thin layer lithography.

The last chapter resumes and compares all technologies to their usefulness in future systems. An appendix shows methods for measuring and qualifying micro-optical components in order to characterize the performance of different elements.

Acknowledgements

We would like to thank Michael Frank for many fruitful discussions.

We are grateful to many collegues for providing us photographic material: Jochen Bähr, Bart Dhoedt, Peter de Dobbelaer, Jost Göttert, Andrew Kirk, Sylvain Lazare, Jörg Moisel, André Müller, Nancy Nieuborg and Stefan Sinzinger.

A very special thank goes to Lucia Feilner for the cover photo.

We greatly appreciate the help of Andy Kirk and Mike Hutley for reading the manuscript and providing us with many helpful suggestions.

We are especially indebted to Irina Veretennicoff and Hugo Thienpont for their support throughout the time we worked for this book.

Maria Kufner, Stefan Kufner
Brussels, August 1996

Transition from Macro-optics to Micro-optics

The most straightforward approach to the miniaturization of optical system is to ask the question: What happens when each conventional optical component is shrinked? To check which properties will be affected in this case the scaling behaviour of the most elementary optical elements, namely lenses and prisms will be considered in this chapter.

1.1. Scaling behaviour of lenses

An example taking into account the different properties of macro- versus microlenses is shown in the following imaging setup [Lohm89, Ozak94]: An extended object has to be imaged through a spatial filter. The object consists of a number of discrete points (fig. 1.1). Then imaging can be performed either by a macro-lens a field of view which covers the entire object (a) or by a set of micro-lenses providing a separate channel for the area of each object point (b).

Fig. 1.1 – Macrolenses versus Microlenses

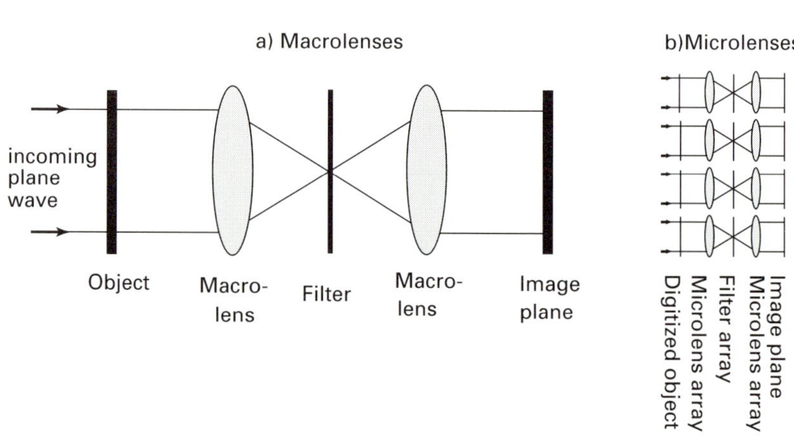

If the system is transformed from a) to b) by scaling diameter and focal length of the lenses by a factor of M all angles are invariant. Correspondingly terms like resolution and depth of focus which depend only on wavelength and aperture angle are invariant, too. Distances scale linear with the factor M, therefore the time of light through the system scales linearly. Moreover this means that surfaces such as the image field size scale as M^2, system volume and weight as M^3.

Furthermore it can be noted that there is one more degree of freedom in the microsystem: In the macro-approach the filter is invariant for the whole object, whereas in the micro-approach there are separate filters for each position which allows space variant filtering.

The following table summarizes the scaling of the most elementary parameters.

Table 1.1 – Scaling of system parameters

PARAMETER	SCALES TO	
lens diameter d	$d \rightarrow M * d$	(1.1)
focal length f	$f \rightarrow M * f$	(1.2)
all angles	invariant	
resolution $\delta x = \lambda F$	invariant	
depth of focus $\delta z = 4\lambda F^2$	invariant	
lateral aberrations ξ	$\xi \rightarrow M * \xi$	(1.3)
travel time of light Δt	$\Delta t \rightarrow M * \Delta t$	(1.4)
image field size $A_F = \Delta x \Delta y$	$\Delta x \Delta y \rightarrow M^2 * \Delta x \Delta y$	(1.5)
system volume V	$V \rightarrow M^3 * V$	(1.6)
system weight W	$W \rightarrow M^3 * W$	(1.7)

with the wavelength λ, the stop number F (which is proportional to the aperture angle of the lenses) and Δx, Δy the lateral dimensions of the image field.

An important measure in the context of information processing is the space-bandwidth-product which represents the number of resolvable points in the image plane or respectively the number of data channels which can be handled in parallel.

It is defined by

$$SW = \frac{A_F}{A_P} \qquad (1.8)$$

with SW denoting the space-bandwith-product, A_F the field size and A_P the area of one pixel. Its scaling can now be derived from the scaling of the above terms.

The size of one pixel in the image plane is calculated from the resolution δx and the lateral aberration $\delta\xi$ of the lens by

$$A_P = (\delta x)^2 + (\delta\xi)^2 \qquad (1.9)$$

With the scaling of δx and $\delta\xi$ as indicated in the table above the pixel size A_P scales to

$$A_P = (\delta x)^2 + (\delta\xi)^2 \rightarrow (\delta x)^2 + M^2 * (\delta\xi)^2 \qquad (1.10)$$

Together with the scaling of the image field size

$$A_F = \Delta x \Delta y \rightarrow M^2 * \Delta x \Delta y \qquad (1.11)$$

the scaling of the space-bandwidth-product follows from its definition:

$$SW = \frac{A_F}{A_P} \rightarrow \frac{M^2 * \Delta x \Delta y}{(\delta x)^2 + M^2 * (\delta\xi)^2} \qquad (1.12)$$

Thus for a lens system without any aberration ($\delta\xi = 0$) the space-band-with-product scales with M^2

$$SW \rightarrow M^2 * SW \quad (\text{for } \delta\xi = 0), \qquad (1.13)$$

whereas for a system where the diffraction effects are neglible compared to the aberrations ($\delta x \ll \delta\xi$) the space-bandwith-products rests unchanged.

$$SW \rightarrow SW \quad (\text{for } \delta x \ll \delta\xi) \qquad (1.14)$$

Particularly for the transition from macro- to microlenses this means that there are principle changes in the scaling laws as soon as the lenses are so small that diffraction effects become significant. This is the case when the effects of aberrations become less important compared to the larger Airy disc resulting from diffraction at the aperture stop.

For a high space-bandwith-product a focus spot size as small as possible would be desirable. The spot enlargement from lateral aberrations is decreased when the system is scaled down. But at the same time a diffraction blur is introduced. For this reason a balance has to be found between the spread from aberration and diffraction so that the focus spot is as small as possible. The question is: how to find an aperture stop which yields the minimum spot size for a given microlens?
An iterative algorithm to calculate from a given lens shape the optimum size of an aperture stop which minimizes the focus spot size is presented in [Test95]. The algorithm is based on Rayleigh's quarter-wave criterion.

Fig. 1.2 – Classification in diffractive and refractive optical elements

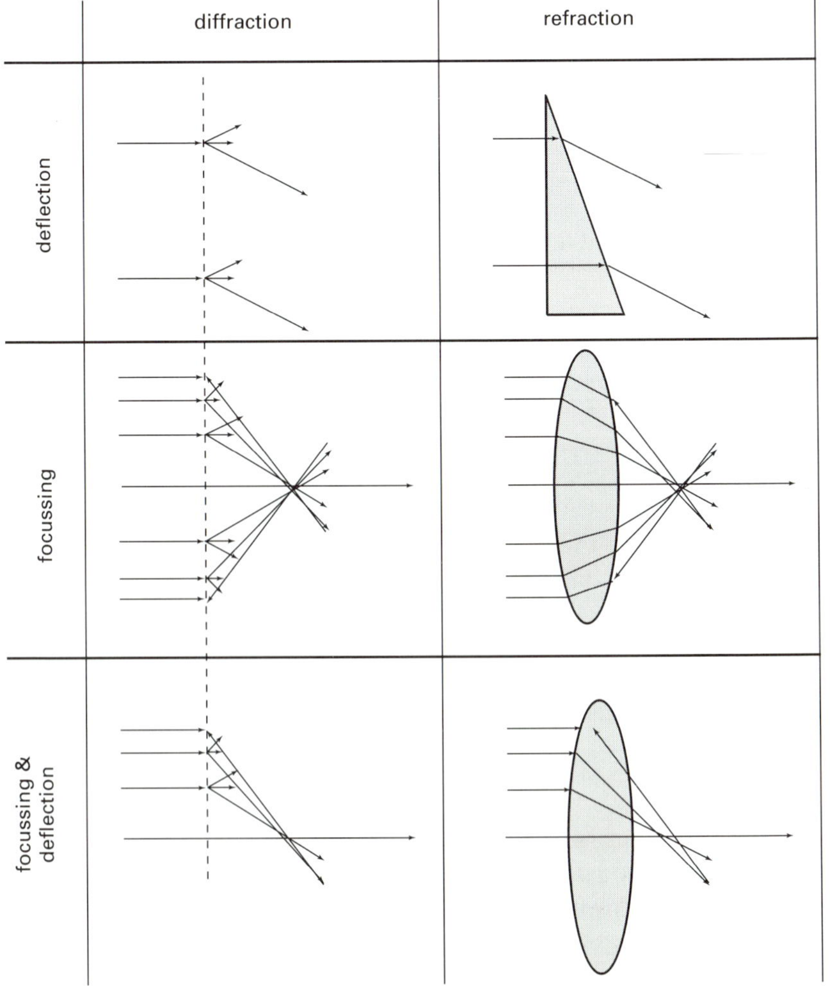

1.2. Transition between diffractive and refractive optics

Consequently the classification into refractive or diffractive optical elements which is used in macro-optics (e.g. fig. 1.2.) cannot be directly transferred in the sense of two counterparts, in these ranges of magnitude there is a smooth transition between refractive and diffractive effects [Sinz95] .

The influence of diffraction at the aperture stop of a small refractive lens is one example for elements in-between refraction and diffraction. Another example is the Talbot self-imaging effect occuring at refractive microlens arrays: The periodic phase pattern is reproduced by diffraction [Lohm90].

For the case of beam deflection it is obvious that there can be intermediate steps between a blazed grating and a prism (fig. 1.3):

Fig. 1.3 – Transition between diffractive and refractive optical elements

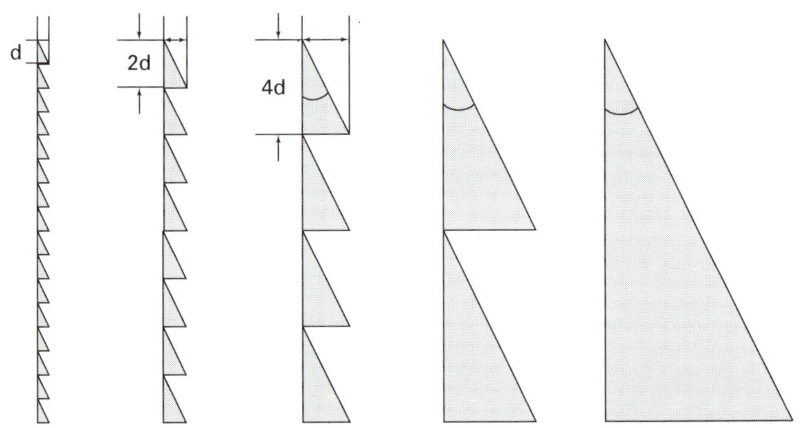

The intermediate versions are created just by keeping the tilt angle and introducing phase steps in multiples of the wavelength. On the one extreme there are phase steps of one wavelength, on the other extreme there are no steps at all. (The same model can be applied in analogy to the transition from a Fresnel-zone-plate to a purely refractive lens.)

The design of phase steps in multiples of the wavelength indicates that there is a high sensitivity to the wavelength in the diffractive approach. For the grating there is a chromatic dispersion caused by diffraction (microdispersion), for the prism there is a chromatic dispersion caused by the prism material (macrodispersion). The two effects are in reverse order, and in practice wavelength sensitivity is much lower in the refractive case, as it is suggested in the example in fig. 1.4.

Fig. 1.4 – Macrodispersion versus microdispersion

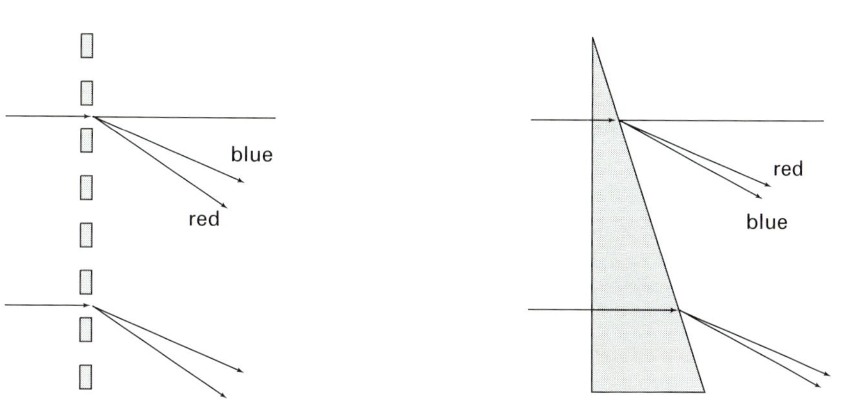

The reverse order of these two effects may suggest a search for an intermediate level where the microdispersion compensates the macrodispersion. However, in blazed periodic components such as shown in fig. 1.3. diffractive and refractive dispersion do not compensate (as it is possible for refractive-diffractive doublets), except for two distinct wavelengths. The design of optical blaze depth then determines for which two wavelengths the compensation works.

1.3. Limitiations for miniaturization

1.3.1. Technological Constraint: Lithographic resolution in the diffraction gratings

There are technical and physical limitations for miniaturization [Smit94, Hutl95]:
The minimum feature size of the fabrication process for diffraction gratings limits the maximum possible deflection angle α of the gratings by

$$\alpha < \arcsin \frac{\lambda}{p_{min}} \qquad (1.15)$$

for a wavelength λ and a minimum grating period of p_{min}. For a n-level etched grating the minimum grating period is correspondingly the n-th multiple of the best resolution r, then the angle α can be expressed by

$$\alpha < \arcsin \frac{\lambda}{nr} \qquad (1.16)$$

Assuming now a system configuration where for example an input array has to be connected in any arbitrary permutation to an output array, as it is depicted in fig. 1.5. the angular constraint limits the miniaturization of the whole system for the following reason:

Fig. 1.5 – Minimum system length caused by a maximum deflection angle

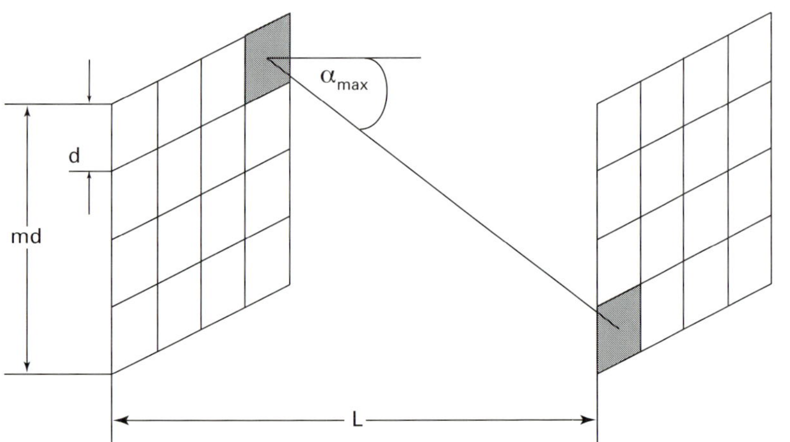

The maximum deflection angle α_{max} required to connect those two channels which are the farest away from each other may not exceed the maximum possible deflection angle α of the grating. In the case of a quadratic array of m by m elements with a distance of d the maximum angle α_{max} occurs for a diagonal connection and is

$$\alpha_{max} = \arctan \frac{\sqrt{2}\,(m-1)\,d}{L} \qquad (1.17)$$

with L denoting the system length, which is here the distance between input and output array. From these two angular restrictions it follows that for a given number of elements and a given resolution the system length cannot be arbitrarily small. Combing equations 1.16. and 1.17. and using the small-angle approximation $\alpha = \sin\alpha = \tan\alpha$, the constraint for the system length is

$$L \geq \frac{\sqrt{2}\,(m - 1)\,dnr}{\lambda} \tag{1.18}$$

For example for a resolution of 1μm, a four-level grating, a wavelength of 633 nm and a pitch of 250 μm the system length must be more than 33.5 mm to enable a permutation of 16 x 16 elements.

Fig. 1.6 – Maximum system length caused by the beam size

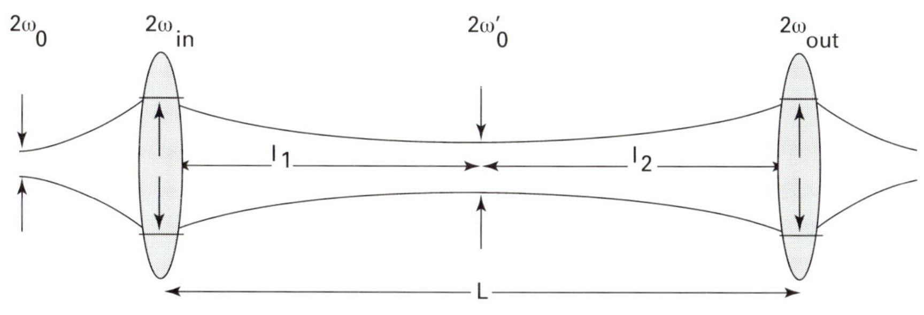

1.3.2. Physical constraint: Gaussian beam shape

Another constraint comes from the fact that a collimated beam of light cannot propagate in free space without increasing in width. Assuming a configuration as depicted in fig. 1.6. where a beam of light enters and leaves the system through a lens, there is a limitation on the distance the light can travel in between these two lenses without exceeding a given width. For an incident beam of light with a Gaussian intensity profile and an incident beam waist of $2\omega_0$ the beam waist in a distance z is given by

$$\omega(z) = \omega_0 \left[1 + \left(\frac{z}{z_r}\right)^2 \right] \tag{1.19}$$

with the Rayleigh length z_r defined as

$$z_r = \frac{\pi}{\lambda} \omega_0^2 \qquad (1.20)$$

Allowing an input beam radius of ω_{in} at the entrance lens and an output beam radius of

$$\omega_{out} = a\,\omega_{in} \qquad (1.21)$$

at the exit lens there is a minimum beam radius ω_0' in between and the system length L can be calculated from the distance l_1 from the input lens to the beam waist and the distance l_2 from the waist to the output lens. Applying the formula (1.19.) to the propagation over the distances l_1 and l_2, the overall system length L can be calculated by [Saue94]

$$L = l_1 + l_2 = \frac{\pi}{\lambda} \omega_0' \left(\sqrt{\omega_{in}^2 - \omega_0'^2} + \sqrt{a^2 \omega_{in}^2 - \omega_0'^2} \right) \qquad (1.22)$$

The maximum propagation length is achieved when the derivative of L is zero:

$$\frac{\partial L}{\partial \omega_0'} = 0 \qquad (1.23)$$

This gives the optimum relation between the beam radius ω_{in} at the entrance and the beam waist ω_0' :

$$\omega_0' = \omega_{in} \sqrt{\frac{a^2}{1 + a^2}} \qquad (1.24)$$

The corresponding maximum propagation length is:

$$L_{max} = \frac{\pi}{\lambda} a\,\omega_{in}^2 = \frac{\pi}{\lambda} \omega_{in}\,\omega_{out} \qquad (1.25)$$

From the point of system scaling this quadratic relation between beam radius and system length means that when the beam size is doubled in lateral direction the system can be four times longer in longitudinal direction.

On the other hand it follows that if a certain system length is required (as it is the case in the previous example) the pitch between two adjacent elements cannot be arbitrarily small, and this limits again the miniaturization of the whole system.

Energy losses, Crosstalk

Tolerances for lateral and longitudinal misalignment are scaled correspondingly with the system size. These errors in the mechanical alignment produce energy losses and/or crosstalk to the nearest neighbours. The amount depends on the particular system configuration, e.g. making the detector size in the output array smaller suppresses crosstalk, but it is paid by energy loss and vice versa.

References for chapter 1

[Hutl95]   M. Hutley: "Case for refracting rather than diffracting optics in spatially variant interconnections", Applied Optics, Vol. 34, No. 14, P. 2448 - 2451, 1995.

[Lohm89]   A. W: Lohmann: "Scaling laws for lens systems", Applied Optics, Vol. 28, No. 23, p. 4996 - 4998, 1989.

[Lohm90]   A. W: Lohmann and J. A. Thomas: "Making an array illuminator based on the Talbot effect", Applied Optics, Vol. 29, p. 4337, 1990.

[Ozak94]   H. M. Ozaktas, H. Urey, and A. W. Lohmann: "Scaling of diffractive and refractive lenses for optical computing and interconnections", Applied Optics, Vol. 33, No. 17, p. 3782 - 3789, 1994.

[Saue94]   F. Sauer, J. Jahns, C. R. Nijander, A. Y. Feldblum, and W. P. Townsend: "Refractive-diffractive micro-optics for permutation interconnects", Optical Engineering, Vol. 33, No. 5, p. 1550 - 1560, 1994.

[Sinz95]   S. Sinzinger and M. Testorf: "Transition between diffractive and refractive micro-optical components", Applied Optics, Vol. 34, No. 26, p. 5970 - 5976, 1995.

[Smit94]   D. E. Smith, M. J. Murdocca, and T. W. Stone in 'Optical computing hardware', chapter 8, "Parallel optical interconnects", p. 193 - 201, ed. J. Jahns and S. H. Lee, Academic Press, Boston, 1994.

[Test95]   M. Testorf and S. Sinzinger: "Evaluation of microlens properties in the presence of high spherical aberration", Applied Optics, Vol. 34, No. 28, p. 6431 - 6437, 1995.

Basics of Lithography

From the very beginning of electronics it was clear that only integration and miniaturization could bring more powerful machines. This lead to an extensive research in lithography as a mean for less expensive mass fabrication of integrated curcuits. These are built up from various arrangements of transistors, diodes, capacitors and resistors, that are individually constructed on a flat silicon or gallium arsenide substrate. For optical applications also glass and polymers are of interest. The patterns defining these regions must first be drawn by lithographic process on a layer of resist material, and then transferred onto the substrate by an etching process. In this context, the lithographic process is the art of making precise designs on thin films of resist material by exposing them to a suitable radiation such as ultraviolet, X-ray, electron or ion beam. Several steps are involved in the lithographic process as shown in fig. 2.1 (see eg. [Cowi91]. The details of each steps are explained each in a subchapter.

- spin coating

A solution of polymer resist is spun onto the substrate surface and baked to remove the solvent and to form a thin film. Typical film thickness for integrated curcuit fabrication is in the range of 0.5 to 2 μm. Optical applications often require thicker films with several tens or even several hundreds of microns.

- masking (optional)

Electromagnetic irradiation uses patterned chromium masks on quartz. Particle irradiation can write the pattern directly onto the resist by a focused small spot.

- exposure

The resist is a polymer sensitive to the radiation used whose chemical or physical properties can be changed by exposure. Depending on the nature of irradiation source and the polymer resist the exposed regions are either rendered soluble if the polymer is degraded (this is called a positive resist) or insoluble if the polymer is crosslinked (this is called a negative resist).

Fig. 2.1 – Steps in the lithographic process for positive and negative resists

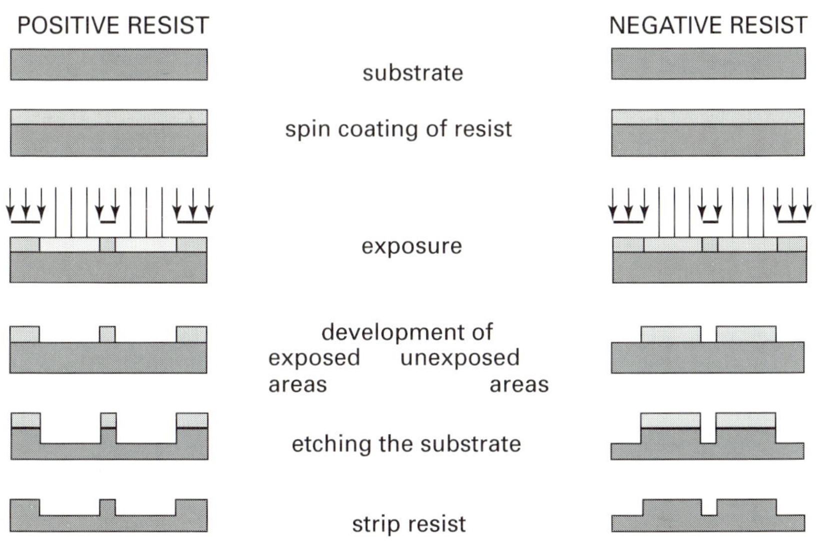

• development

The latent image formed by radiation on the resist can subsequently be developed by treatment with solvents or plasma. After exposure a developer removes either the irradiated regions (positive resist) or the nonirradiated (negative resist).

• etch

Then etching transfers the latent image in the resist to the substrate either by wet etching or by plasma etching. The resist must also be resistant (hence the name) after development to the etchant, and protect the areas still covered with resist. All other regions are attacked and a certain pattern is transferred onto the substrate.

• strip

Once the pattern is in the substrate material the remaining resist is stripped off.

These processing steps can be repeated if several depths of structering have to be achieved. But each time the mask has to be realigned with the help of alignment patterns.

2.1. Spin coating

In the first step a layer of photoresist is spun on a substrate (eg. glass). The layer thickness primary depends on the resist material itself and is a function of spinning time and speed. For one example of a relatively thick resist (AZ 4620-A from Shipley [Hoec89]) some data are listed: fig. 2.2 shows the correlation between the layer thickness and the spinning time for a speed of 2000 rotations per minute (RPM). Table 2.1 shows typical results for varying spinning time for the same resist [Daly90]. Homogeneous film thickness is limited to 30 μm maximum. Thicker films can be fabricated by repeated spinning of several layers on one substrate [Völk95] (see chapter 7).

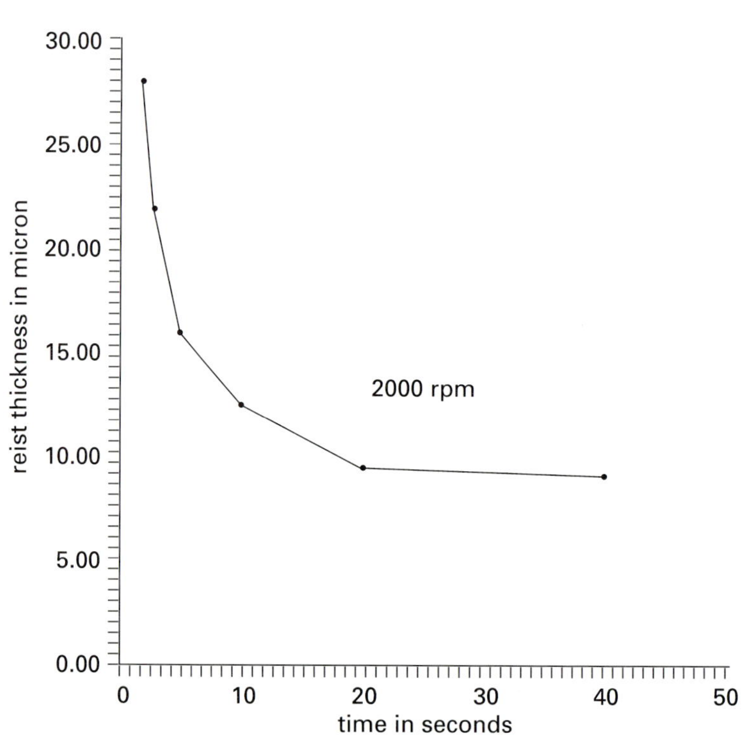

Fig. 2.2 – Example of photoresist layer thickness as a function of spinning time (After [Hoec89])

Table 2.1

RPM	time/s	Thickness/μm
2500	10	8
1200	10	20
800	6	40
800	4	50

2.2. Masking

The resist can be patterned by using a mask and exposing it to a homogenous radiation source, or the designed pattern can be directly written into the resist. Different methods exist for the mask fabrication as well as for the direct writing.

2.2.1. Mask fabrication

- demagnification of plotter image

A first approach to mask fabrication is to design a black and white pattern on the computer and print it with a plotter in a magnified format (factor 10 to 30). Then a high quality camera system demagnifies the plot onto a high resolution plate. The developed plate serves as mask for replication into photoresist.

- direct writing of chromium masks

Direct writing of the mask results in better mimimum feature sizes. Here a beam is scanned in a raster over a substrate (typically quartz glass covered with a chromium layer and a photoresist layer on top of it). At each adressable position in the raster field, which is determined by the spot size of the beam, the resist can be irradiated. Laser scanners can hardly go under 1 μm line width. Electron beam writing however offers better resolution. Beam sizes can be the order of 10 nm. This has become the standard technique for mask fabrication. Typical dimensions of a chromium mask on quartz are 10 by 10 cm^2. The minimum feature size can be about 0.35 μm down to 0.25 μm [Lang95]. If even better resolution is required, the electron beam can be replaced by an ion beam. Then a line width of 0.1 μm is possible (see also chapter 3.3).

2.2.2. Transfer of a pattern to the resist

- contact printing

In contact printing the mask is pressed against the resist. This is the best transfer method of the mask pattern into the resist with respect to diffraction. A big disadvantage is the damage of the rather expensive mask after only a few irradiation cicles. A minimum feature size of 0.3 μm with an irradiation wavelength of 250 nm has been reported [Watt87].

- proximity printing

The disadvantage of mask damage does not occur in proximity printing when the mask has a distance of a few micrometers from the substrate. Typically the distance lies between 20 μm and 50 μm. Due to the close distance diffraction effects are reduced compared to projection printing. It is clear that a shorter wavelength gives reduced diffraction. This is also a reason for deep UV irradiation in this technique. Many production lines work with this technique and the standard minimum feature size for 1995 is 0.35 μm. But a further improvement is possible by using X-ray lithography. There diffraction can be neglected. Nevertheless it has to be mentioned that deep UV is cheaper as long as it can provide the same structural line width and it can use the standard chromium mask on quartz. X-ray lithography however has a need for more complicated mask design (see chapter 9).

- projection printing

Instead of demagnifying a plotter image onto a high resolution plate it can directly be written into the resist of the later IC by projection printing. Due to imaging errors a minimum feature size of less than 2 μm is hard to achieve using UV light of 300 nm and above. But in recent years excimer laser with a wavelength of 190 nm are replacing older UV techniques. Difficulties arise because of the inhomogeneous beam profile of the laser. Sophisticated quartz optics has to be designed for a successful use in a production line. Minimum feature size of less than 0.5 μm has been reported [Watt87].

2.3. Exposure

The exposure of resists to any type of radiation can change the chemical structure of the resist material. As illustrated in fig. 2.1 resists are classified positive or negative. Positive means degradation of the polymer and negative means crosslinking of the polymer. In most cases both reactions occur simultaneously and the term positive or negative resist

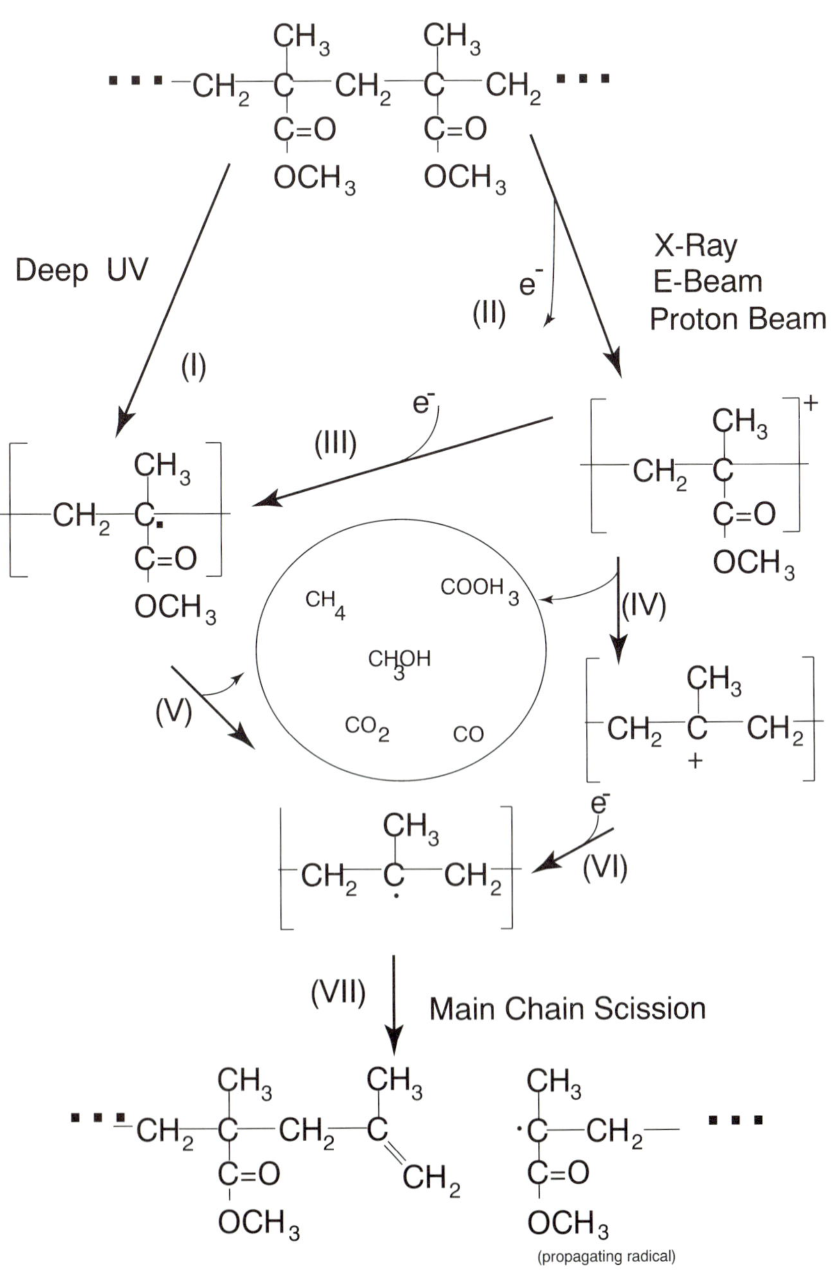

Fig. 2.3 – Degradation of PMMA by various radiation sources
(After [Choi88])

only indicates what reaction dominates. A typical positive resist is poly (methyl methacrylate) (PMMA). It is one of the oldest known resists and is also one of the most widely used. PMMA is a very slow resist meaning that there exist many other resists which are much more sensitive to irradiation (see table 2.2.) . But nevertheless it has shown line resolution down to 50 nm. Another important feature is its sensivity to all irradiation sources [Choi88]. This is not a usual fact. In general resists are sensitive only to one irradiation source. Fig. 2.3. shows main chain scission in PMMA by electromagnetic (UV and synchrotron irradiation) and by particle irradiation (electron and proton irradiation).

Table 2.2. summarizes sensivity data of positive and negative resists and gives also a figure for the achievable contrast ($[\log(100\%) - \log(0\%)]^{-1}$ in terms of intensity seen by the resist for irradiation with an electron beam [Watt87, Peth91].

Table 2.2

RESIST	POL	SENSIVITY (C/cm^2) @ 20kV	Resolution [μm]	Contrast
PBS (Mead Tech.)	+	1.8×10^{-6}	0.5	1.7
PMMA (KTI Chem.)	+	1×10^{-4}	<0.1	2
EBR-9 (Toray Ind.)	+	1.2×10^{-6}	0.5	3
FBM-110 (Daikin Ind.)	+	1.5×10^{-6}	1.5	5
AZ 2400 (Shipley Co.)	+	2×10^{-4}	0.5	2
COP (Mead. Tech.)	-	5×10^{-7}	1.5	0.8
OEBR-100 (Tokyo Okha)	-	5×10^{-7}	1.5	0.8
SEL-N (Somar Ind.)	-	1×10^{-6}	1	0.6
GMCIA (AT&T)	-	7×10^{-6}	0.5	1.7
CMS (Toyo Soda)	-	2×10^{-6}	0.7	1.5
RE-4000 N (Hitachi Chem.)	-	3.5×10^{-6}	1	1.3

A quantitative analysis of the irradiation effects in polymers has been given by Charlesby [Char91]. It is assumed that the chemical process associated with the irradiation is completely random and that the polymer is linear (without crosslinks).

The dose given to the polymer is expressed in units of a Sievert (Sv), defined as:

$$1\ Sv = 10^{-5}\ \frac{J}{g} = 0.625 * 10^{14}\ \frac{eV}{g} \tag{2.1}$$

A measure for the effect of radiation to a polymer is the so-called G-value which depends strongly on the type of irradiation source. For the case of a positive resist it is defined as the number of chain scissions per 100 eV absorbed energy. Using this definition a dose D causes N scissions with

$$N = D * \frac{G}{100\ eV} \tag{2.2}$$

The change of the average molecular weight of a polymer from an initial molecular weight M_0 to a molecular weight M_D after absorption of a dose D can then be calculated as follows: The initial number of molecules per gram is given by N_A/M_0. The number of molecules after absorbtion of a dose D is given by N_A/M_D. N_A is the Avogadro number. Since every scission increases the number of molecules by unity, the total change in the number of molecules is:

$$\frac{N_A}{M_D} = \frac{N_A}{M_0} + D * \frac{G}{100\ eV} \tag{2.3}$$

From this follows directly the change of the inverse molecular weight from irradiation with a dose D:

$$\frac{1}{M_D} = \frac{1}{M_0} + \frac{D}{N_A} * \frac{G}{100\ eV} \tag{2.4}$$

In the literature almost for every combination of polymer and radiation its G-value can be found. G-values do not only exist for degradation but also for crosslinking which will not be discussed here. Historically the G-value was introduced for the characterization of thin resist layers where the linear behaviour of equation (2.4) is valid. There a G-value independant of dose occurs. Later on it will be shown that thick resists have other functionality because of higher dose. Table 2.3 gives some typical G-values with γ-radiation [Reis89].

Table 2.3

Polymer	G-value
PMMA	1.2
Poly(isobutylene)	4.0
Poly(butene-1-sulfone)	8.0

More features which are of practical importance for the polymer resists are:
- reduced swelling during development
- good adhesion with the substrate material
- high photosensibility
- high contrast
- high resolution
- chemically and physically resistant against the etchant
- easy strip off

2.4. Development

Commercially available resists consist of three components:
- a photoactive compound,
- a base resin and
- a solvent.

The base resin is soluble in the developer (normally aqueous alkaline developers), and the presence of the photoactive compound strongly inhibits its dissolution. Irradiation neutralizes the photoactive compound and increases solubility (in case of a positive resist). For most applications the slope of both the resist structures and the etched substrate structures have to be known. In many cases the resist development is better known experimentally rather than its mathematical treatment because of strong nonlinearities of the processes. The first step is to determine the solubility rate (V) as a function of dose (D). Resist areas are irradiated with different known doses and the etch rate of the developer in the resist is measured as a function of development time. A typical result for the resist AZ-1350 is shown in fig. 2.4 [Mell95].

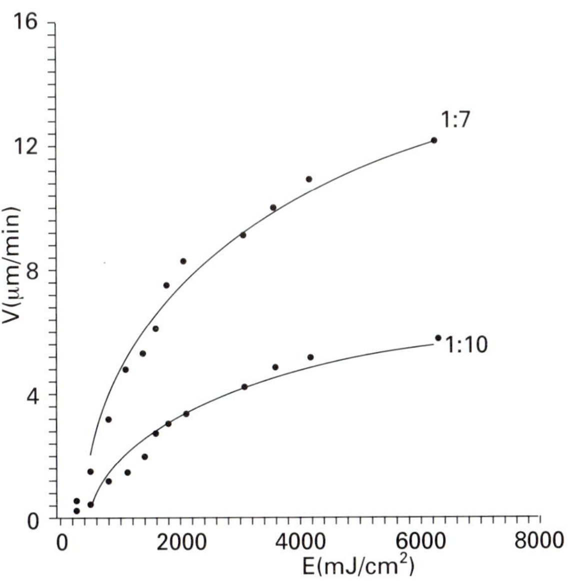

Fig. 2.4 – Solubility rate V as a function of irradiation dose D
(After [Mell95])

For more complex structures the intensity distribution in the resist has to be known. Then the development process is described by the following differential equation

$$\frac{\delta \bar{r}}{\delta t} = -V(\bar{r}) * \bar{n} \tag{2.5}$$

where

$$\bar{n} = \sin(\phi)*\bar{x} + \cos(\phi)*\bar{z} \tag{2.6}$$

is the unit vector normal to the surface in x and z direction. Introducing discrete time steps equation (2.5) transforms to

$$\bar{r}(t+\Delta t) = \bar{r}(t) - V(\bar{r}(t))*\Delta t*(\sin(\phi)*\bar{x} + \cos(\phi)\bar{z}) \tag{2.7}$$

In literature several reasons for the effects that limit the final resolution in every resist are discussed:

- chemical diffusion of photoactive compound
- chemical diffusion of developer

- dependance of development rate on surface curvature
- mean distance between photosensitive molecules
- size of smallest photoresist particle that can be removed by the developer

2.5. Etching

Once the resist pattern is formed on the substrate, it can be etched into the substrate. The pattern in the resist can be transferred into the substrate either by wet etching or by dry etching. In case that the resist layer does not resist the etching process a lift-off technique has to be applied, depicted in fig. 2.5. The resist pattern always shows a certain undercut after development. This influences the lateral accuracy. To avoid this undercut for the subsequent etch procedure a thin metal film is deposited over the substrate by vacuum deposition or sputtering (the resist thickness must be larger than the film thickness). An organic solvent removes the resist and a negative mask compared to the resist mask is now on the substrate.

Fig. 2.5 – Lift-off technique

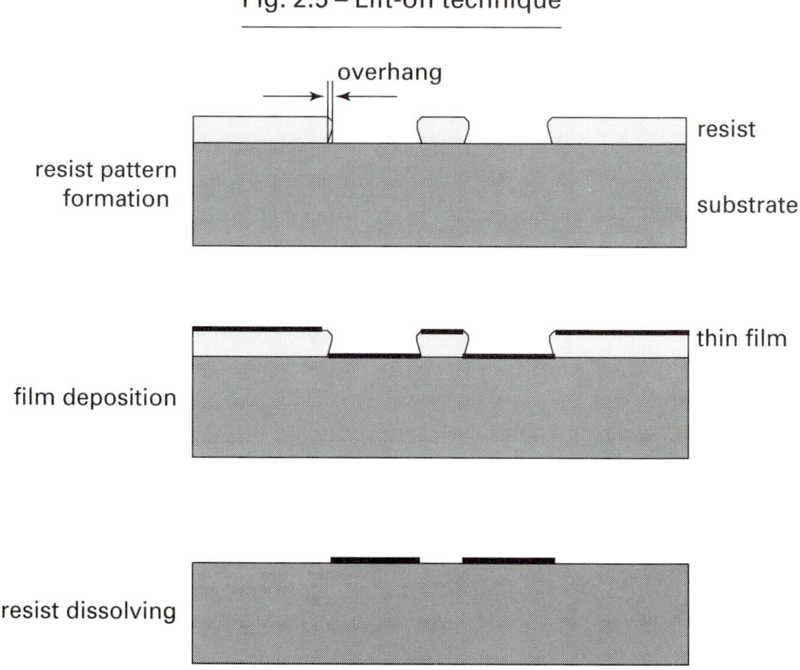

Table 2.4

material	etchant
Al	5-10 % NaOH aqueous solution
Au	I_2 1.2g + NH_4I 3g + H_2O 40cc + C_2H_5OH 60cc
Ag	HNO_3 (3:1 diluted)
Cu	$FeCl_3$ aqueous solution
Ti	HF acid : H_2O = 1:9
Cr	$K_3(Fe(CN)_6)$ 30g + NaOH 5g + H_2O 100cc

Many acids, alkalis and saline solutions are available in wet etching. The specimen is dipped in the etchant that has kept at an appropriate temperature and has been kept stirred to eliminate air bubbles. Wet etching is a cheap solution with easy handling. But good results with minimum feature sizes of less than 1 μm can hardly be expected because the process is isotropic. Very high process control is necessary to provide good reproducibility. Table 2.4. summarizes some etchants for different materials [Nish89].

Dry etching techniques have been developed to overcome the drawbacks of wet etching. Directional, anisotropic, etching is associated with the action of fast moving ions impinging in one direction on the surface of the substrate.

Fig. 2.6 – Reactive Ion Etching chamber

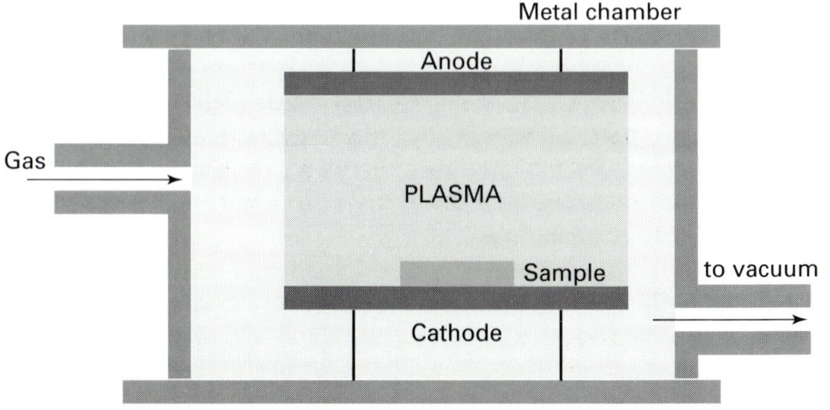

Fig. 2.7 – Anisotropic and isotropic etching

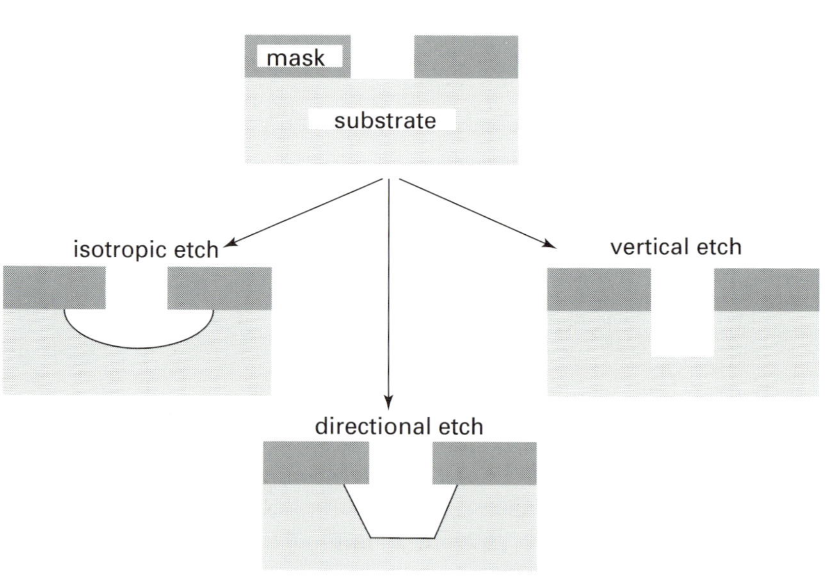

Fig. 2.6. shows an etch chamber where Reactive Ion Etching (RIE) is applied. An etch gas is pumped into the chamber. The electrical field between anode and cathode forms a reactive gas plasma. Positive ions are accelerated toward the sample which is on the same potential as the cathode. The directional ion bombardment in RIE enhances the etch rate in the vertical but not in the horizontal direction and this produces anisotropic etching. The ratio of ions to neutrals reaching a surface and the kinetic energy of the ions determine the directionality of the etching process. Fig. 2.7. depicts schematically the differences in isotropic and anisotropic etching [Ster94].

Table 2.5 shows some typical RIE resists which can resist the etchant better than the substrate does. The rates are given for an oxygen plasma. Selectivity is often used in literature as a synonoym expression for the relative etch rate. The corresponding G-values of the materials are also listed in the table [Reis89].

Table 2.5

Polymer	Etch rate (relative to substrate)	$G_{scission}$
Poly (o-methylstyrene)	1.11	0.3
Poly (phenyl methacrylate)	1.33	-
Poly (vinyl methyl ketone)	1.48	-
Poly (methyl methacrylate)	2.37	1.2
Poly (isobutylene)	3.56	4.0
Poly(butene-1-sulfone)	7.11	8.0

Fig. 2.8 – Diffractive microlens pattern etched in GaAs. Focal length: 135 μm, Diameter: 250 μm. (Courtesy Bart Dhoedt, [Dhoe95])

In fig. 2.8 a SEM picture of a diffractive microlens is shown. The lens is etched in GaAs. The focal length of the lens is 135 μm in air and the lens diameter is 250 μm. Smallest features on the lens are 1 μm. Etching was done by RIE with $SiCl_4$.

The etch process strongly depends on the atomic number of the substrate. High Z materials show anisotropic etching, whereas low Z numbers tend to an isotropic etch process. Materials with low Z number are quartz, silicon and most glasses. In the following example SiO_2 is used as substrate which also is a low Z material [Grat93].

Fig. 2.9 – Surface evolution of point h(x,t) to h(x,t+dt), where n is the surface normal, ds is the etch along the normal and dh is its projection along h [Grat93]

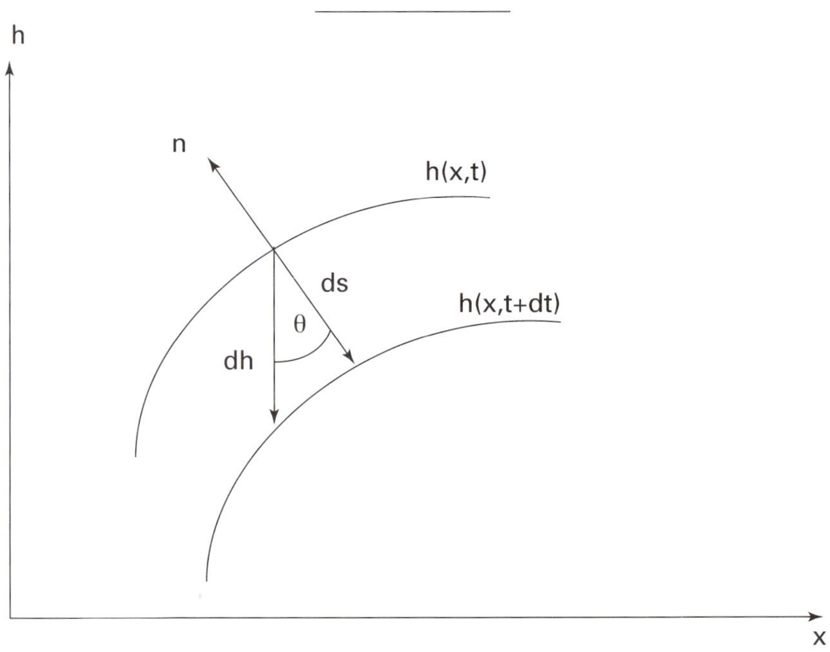

For a surface h(x,t) having been etched into a surface h(x,t+dt), a time dt later, it can be defined a distance ds etched in the direction of the surface normal n̄ and a vertical projection (see fig. 2.9.). It can be shown that the etch rate along the surface normal is R, where

$$\frac{ds}{dt} = R\,\bar{n} \quad \text{with} \quad R = R\left(x,\, t,\, h,\, \frac{dh}{dx},\, \frac{d^2h}{dx^2},\, \ldots\right) \tag{2.8}$$

In the functional dependancy of R the physical properties of the process are incorporated. From fig. 2.9. it can be seen that a vertical etch rate dh/dt can be expressed as

$$\frac{dh}{dt} = \frac{dh}{dt}\,\frac{1}{\cos\theta} \tag{2.9}$$

For the description of an isotropic etch process the etch proceeds equally in all directions. Physically this means that $R_i(\theta)$ is constant for all angles. With equation 2.9. it follows

$$\frac{dh_i(\theta)}{dt} = \frac{\rho}{\cos\theta} \quad \text{with} \quad R_i(\theta) = \rho \tag{2.10}$$

For an anisotropic process $R_a(\theta)$ must vary with a slope in such a way as to cancel out the $1/\cos(\theta)$ effect and the rate of change in height $dh(\theta)/dt$ must be constant everywhere. From the starting condition $R_a(0) = R_i(0) = \rho$ it follows

$$\frac{dh_a(\theta)}{dt} = \rho \quad \text{with} \quad R_a(\theta) = \rho\cos\theta \tag{2.11}$$

Fig. 2.10. shows the etch rate for an ideal anisotropic etch process and a typical ion milling process. The zero angle etch rate was selected as $\rho = 200$ A/min which is representative for resist and SiO_2.

Fig. 2.10 – Etch rates for anisotropic and typical ion mill etching
(After [Grat93])

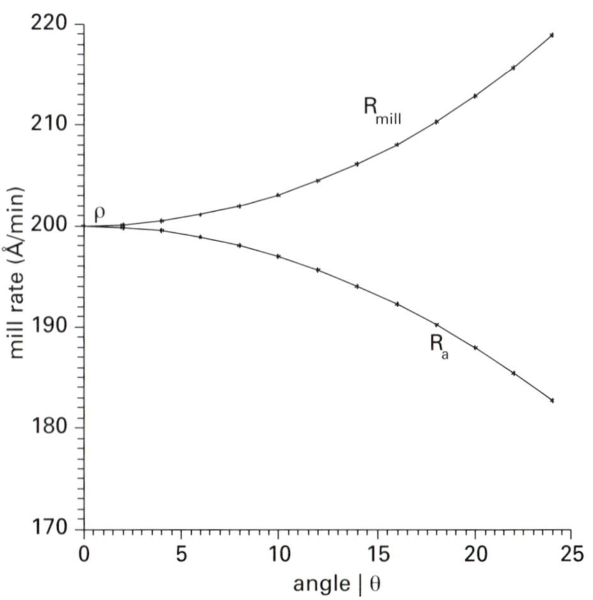

References for chapter 2

[Char91]   A. Charlesby in 'Irradiation effects in polymers', chapter 2, "The effects of ionising radiation on polymers", p. 39 - 78, ed. D. W. Clegg, A. A. Collger, Elsevier, London 1991.

[Choi88]   J. O. Choi, J. A. Moore, J. C. Corelli, J. P. Silverman, and H. Bakhru: "Degradation of poly (methyl methacrylate) by deep ultaviolet, x-ray, electron beam and proton beam irradiating", J. Vac. Sci. Technol. B 6 (6), p 2286 - 2289, 1988.

[Cowi91]   J. M. G. Cowie: "Polymers: chemistry & physics of modern materials", Chapman&Hall, New York 1991.

[Daly90]   D. Daly, R. F. Stevens, M. C. Hutley, N. Davies: "The manufacture of microlenses by melting photoresist", J. MEAS. Sci. Technol. 1, P. 759 -766, 1990.

[Dhoe95]   B. Dhoedt: "Theoretische en Experimentele Studie van Vrije-Ruimte Optische Interconnecties op Basis van Diffractieve Lensrijen", Thesis Uni Gent, 1995.

[Grat93]   E. J. Gratix: "Evolution of a microlens surface under etching conditions", in Miniature and Micro-Optics and Micromechanics, N. C. Gallagher, Jr., C. Roychoudhuri, ed., Proc. SPIE 1992, p. 266 - 274, 1993.

[Hoec89]   Hoechst product information, "AZ 4620: Positiv Fotoresist für hohe Schichtdicken", 1989.

[Lang95]   J. C. Langston and G. T. Dao: "Extending optical lithography to 0.25 mm and below", Solid State Technology, p. 57 - 64, March 1995.

[Mell95]   B. de A. Mello, I. F. da Costa, C. R. A. Lima, and L. Cescato: "Developed profile of holographically exposed photoresist gratings", Applied Optics, Vol. 34, No. 4, p. 597 - 603, 1995.

[Nish89]   H. Nishihara, M. Haruna, and T. Suhara: "Optical integrated circuits", McGraw Hill, New York 1989.

[Peth91]   R. A. Pethrick in 'Irradiation effects in polymers', chapter 10, "Microlithography", p. 383 - 430, ed. D. W. Clegg, A. A. Collger, Elsevier, London 1991.

[Reis89]   A. Reiser: "Photoreactive polymers", Wiley, New York 1989.

[Ster94]   M. B. Stern and T. R. Jay: "Dry etching for coherent refractive microlens arrays", Optical Engineering, Vol. 33, No. 11, p. 3547 - 3551, 1994.

[Völk95]   R. Völkel, P. Nussbaum, K. J. Weible, H. P. Herzig, R. Dändliker, S. Haselbeck, M. Eisner, and J. Schwider: "Fabrication of non-conventional microlens arrays", in Microlens Arrays, Vol. 5, 1995 EOS Topical Meetings Digest Series, p. 116 - 120, 1995.

[Watt87]   R. K. Watts in 'VLSI technology', chapter 4, "Lithography",
p. 141 - 183, ed. S. M. Sze, McGraw Hill, New York 1987.

Diffractive Optics

Diffractive optical elements (DOE) can be fabricated with either analog or digital techniques. Analog DOEs are formed by the optical interference of waves as done for the recording of holograms. In this chapter only digital diffractive optical elements will be discussed, since they directly use the lithographic masks already explained in chapter 2. For the case of miniaturized diffractive optics special interest is focused on the basic optical functions: light deflection and light collimation.

3.1. Classification of gratings

3.1.1. Binary gratings

The most straightforward approach to a diffraction grating is a binary periodic structure with altering tranparent and opaque regions or altering phase steps as indicated in fig. 3.1.

For such a grating with a period p the first order of diffraction ("+1") appears at an angle α according to formula 3.1.:

$$\sin \alpha = \frac{\lambda}{p} \tag{3.1}$$

for a wavelength λ. On a screen in a distance D from the grating the first diffraction order is displayed in a distance y from the optical axis:

$$\tan \alpha = \frac{y}{D} \tag{3.2}$$

For small angles it follows

$$y = \lambda \, \frac{D}{p} \tag{3.3}$$

Fig. 3.1 – Periodic optical grating

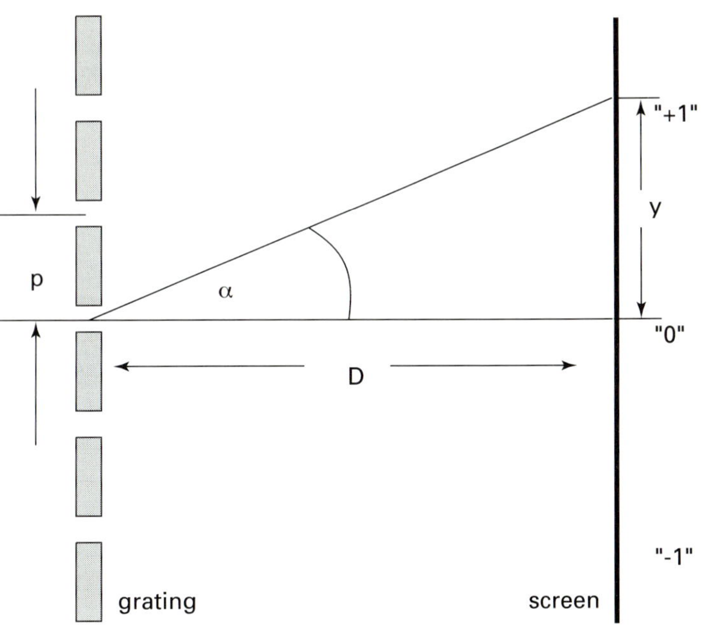

Assuming for example a wavelength of $\lambda = 0.5$ µm and a pitch of p = 50 µm results in an angle of deflection of $\alpha = 0.6°$. To achieve a deflection angle of $\alpha = 5°$ a pitch of 5 µm would be required. In the latter case a displacement of the first order with y = 100 µm would need a distance D = 1.1 mm beween grating and screen. It is clear from this few examples that digital diffractive optics has difficulties in providing large deflection angles and compact systems because it is directly correlated to the limitations in minimum feature size (see also chapter 1.3.1).

3.1.2. Multilevel gratings

A periodic grating such as the one in fig. 3.1 has a very low diffraction efficiency into order "+1". The efficiency improves when the number of phase levels increases and the structure approaches a prism. The quantization to discrete levels is achieved by n masks which results in 2^n possible etch depths. The principle is shown in fig. 3.2 how for n = 3 masks 8 different etch levels can be achieved.

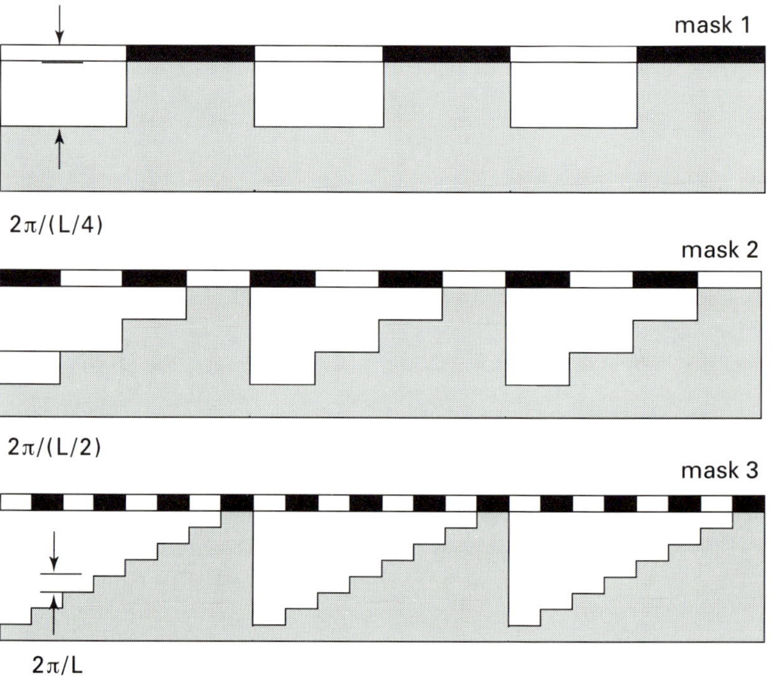

Fig. 3.2 – 2^n different etch levels by use of n masks

The higher n is the more the staircase looks like a blazed grating. For a periodic grating corresponding to the previous example with a pitch of p = 5 μm which gives the same deflection angle of 5° the required resolution r would be:

r = 0.625 μm if L = 8 (n = 3) and r = 1.25 μm if L = 4 (n = 2).

A look at the difference in diffraction efficiency of this two cases is the most interesting aspect in the design. The diffraction efficiency η for order "+1" is given by [Jahn94]

$$\eta_{+1} = \mathrm{sinc}^2(\frac{1}{L}) = \left(\frac{\sin(\frac{\pi}{L})^2}{\frac{\pi}{L}}\right) \tag{3.4}$$

From this formula the results given in table 3.1. are calculated:

Table 3.1

n	$L = 2^n$	η_{+1}
1	2	40.5
2	4	81.0
3	8	94.9
4	16	98.7

3.1.3. Blazed gratings

The relative alignment of the subsequent masks necessary for quantization tends to limit the performance of the final component. There exist several methods for direct writing of blazed structures which use different irradiation sources: laser writing, electron beam writing and ion beam writing. The blazed surface profile that is created in the resist can be transferred into the substrate using wet or dry etching techniques. The substrate material can be glass, metal or plastic. A metal master is useful in case of less expensive replication.

Laser writing techniques

Based on photolithography a laser pattern is generated into a photoresist layer (fig. 3.3 a). Two geometries of laser pattern generation are available. In the first approach (r-θ) the substrate is spun on an air-bearing spindle. This is widely used for large diameter circular lens design. The second approach moves the substrate in a Cartesian plane (x-y). The resist layer is scanned by a focused laser beam from eg. a helium-cadmium laser (λ = 422 nm) combined with an autofocus system [Gale94]. The laser intensity is modulated as required by the optical design and the beam exposes the photosensitive layer to produce a continuous surface-relief pattern. If high numerical aperture objectives are used, eg. NA = 0.9, the minimum feature size can be 0.5 μm.

Fig. 3.3 – Direct blazed structure generation with laser illumination using variable exposure or grey scale pattern

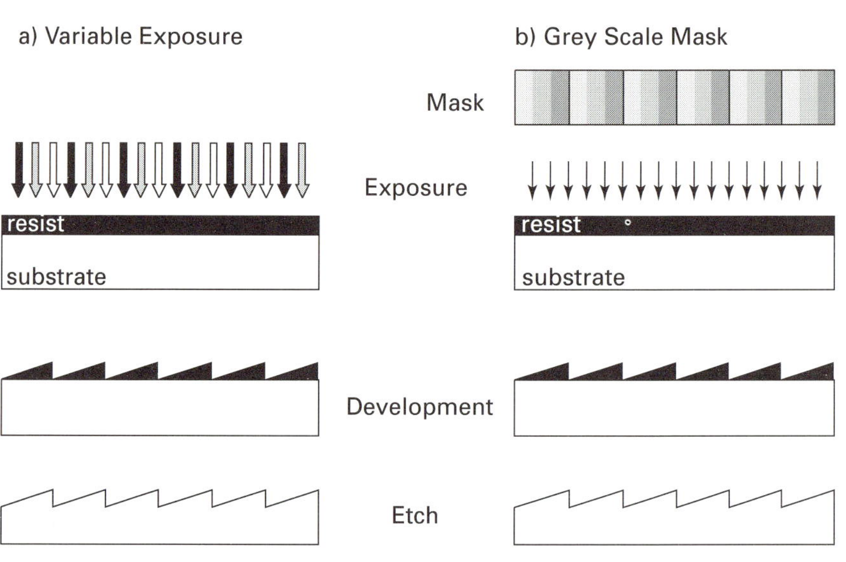

Another laser approach is based on grey scale masks (fig. 3.3 b) [Sule95]. The grey scale mask can be generated by writing a magnified plotter image. The grey scale is obtained by varying the dot density per mm^2. Instead of a plotter also a laser printer with high resolution and desktop computer software can be used to create the grey scale. Then the image is photoreduced on a high resolution plate as described in chapter 2.4. The grey scale is improved by the photoreduction process because of the film response of the high resolution plate. This process is not suitable for minimum feature sizes of less than 2 µm.

Electron-beam writing

For electron-beam irradiation a modified scanning electron microscope (SEM) can be used [Shio94]. The pattern generators, the beam blanking, and the x-y stage are fully controlled by a computer with high processing speed. The beam diameter is typically 10 nm at 35 kV accelerating voltage and 10 pA beam current. Writing the data is the same as already described in chapter 2.4. For the intensity modulation in the resist the speed of the x-y stage determines the time of illumination for every addressable point, whilst the intensity of the beam itself stays constant.

PMMA is often used as resist with a thickness of about 1 μm. Patterns with minimum feature size of 0.1 μm can be fabricated. Typical examples of microlenses show diameters larger than 100 μm and numerical apertures of 0.1. But also numerical apertures of 0.5 have been reported for microlenses with diameters of 600 μm and for a design wavelength of 1.3 μm [Zars94]. Focusing efficiency of better than 70 % have been achieved using 8 design levels of depth. Writing times for small lens arrays are in the order of several minutes. Eg. a 1x 5 array of lenses (125 x 125 μm^2) needs 9 minutes to be written into the resist [Shio94].

Ion-beam writing

Ions are very similar to electrons (see previous paragraph and chapter 2.4). In addition a high ion current in the order of 10^{17} ion/cm^2 leads to sputter effects [Harr94]. A typical system for this application uses a Ga liquid metal ion source and a single electrostatic focusing column. The beam diameter is in the range between 50 nm and 200 nm at 20 to 30 keV. Material removal is accomplished by computer controlled scanning of the ion beam in a x-y Coordinate system. Different dose at each irradiation point is achieved by changing the scanning speed. The sputter process can be improved if reactive gases are in the sputter chamber like eg. XeF$_2$.

3.2. Examples of applications

3.2.1. Fresnel zone plate

A refractive optical element can be transformed into a blazed grating by a subtraction of multiples of the wavelength λ from the optical path. Fig. 3.4 shows the design principle for the transfer of a refractive to a so-called Fresnel lens. The refractive lens on the left side is sliced in layers of λ. The Fresnel lens on the right side is composed of the remainder modulo λ of each slice and carries (for the design wavelength) the same phase information as the refractive lens.

The calculation of the focal length of a Fresnel lens is illustrated in fig. 3.5. The light rays deflected by ring r_j must arrive in the focal point of the lens. Their optical path length is f+jλ. It follows

$$r_j^2 + f^2 = (f + j\lambda)^2 \tag{3.5}$$

and

Diffractive Optics

$$r_j^2 = 2\,j\,\lambda\,f + (j\,\lambda)^2 \qquad (3.6)$$

With the paraxial approximation ($f \gg j_{max}\lambda$) it follows

$$r_j^2 = 2\,j\,\lambda\,f \qquad (3.7)$$

and

$$f = \frac{r_1^2}{2\lambda} \qquad (3.8)$$

Fig. 3.4 – Transfer from a refractive lens to a diffractive lens
(Fresnel lens)

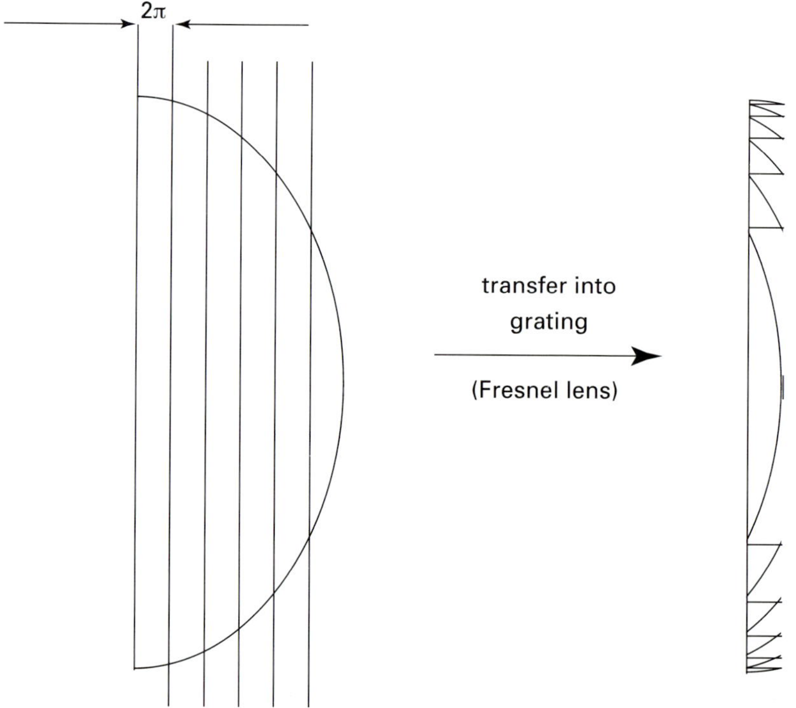

Fig. 3.5 – Calculation of focal length of a Fresnel lens

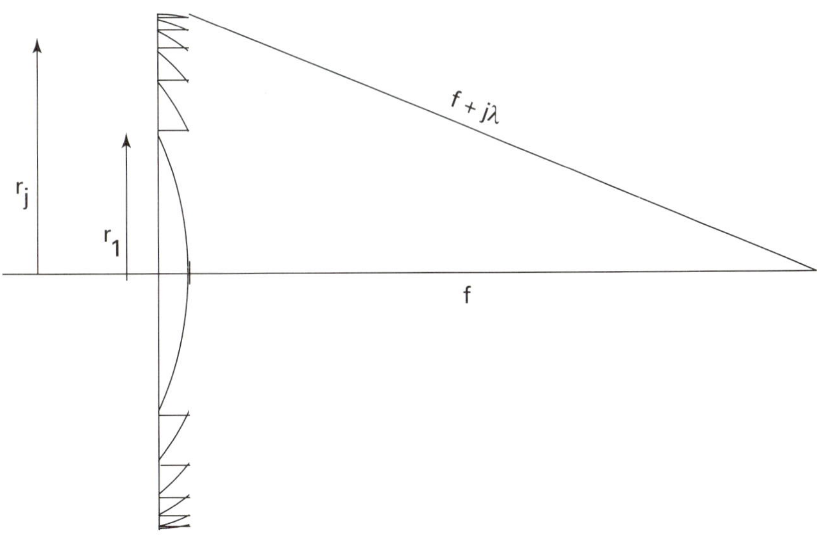

Assuming a wavelength of $\lambda = 500$ nm and $r_1 = 50$ μm it follows for $f = 2.5$ mm. If $r_1 = 10$ μm than $f = 100$ μm. This two simple examples show that the F-number of a lens, defined as the ratio between focal length and lens diameter is restricted by manufacturing limits. Therefore the lenses have long focal lengths in comparison with the diameter. This results in lenses with high F-numbers. If the F-number has to be small then diffraction efficiency suffers, since the number of rings is also limited by lithography. Fig. 3.6 shows a Fresnel lens with a diameter of 250 μm, a focal length of 135 μm in air and a minimum feature size of 1 μm.

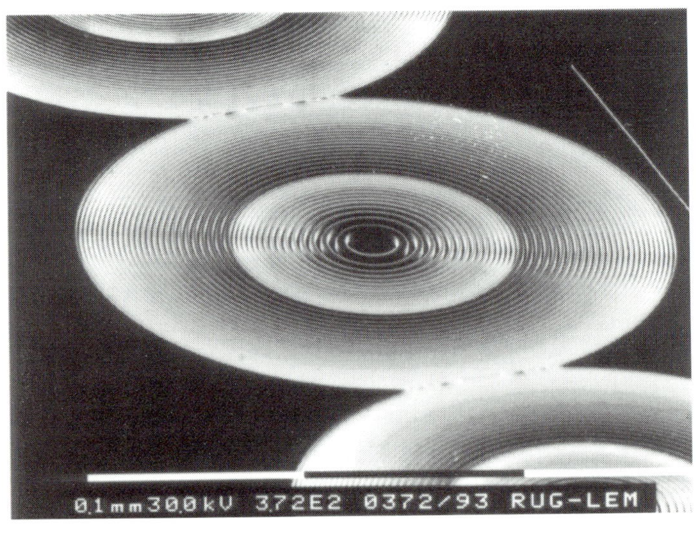

Fig. 3.6 – Diffractive microlens in GaAs. Focal length: 135 μm, Diameter: 250 μm (Courtesy Bart Dhoedt, [Dhoe95]).

The width of the zone levels can be calculated as

$$(\sqrt{N} - \sqrt{N-1})\, r_j \tag{3.9}$$

Using the above value for $r_1 = 10$ μm it follows eg. $r_{10} = 1.6$ μm and $r_{20} = 1.1$ μm. If multilevel etching has to be applied in order to yield higher efficiency the zone width scaling is

$$(\sqrt{N} - \sqrt{N-1})\, \frac{r_j}{L} \tag{3.10}$$

Given by this scaling and from lithographic limitations a compromize is necessary for the outer rings with lateral dimensions < 20 μm. Very often this rings are written with only 4 phase levels (or even binary) whereas the inner zones have 8 or 16 phase levels [Gale94]. Commercially available diffractive microlens arrays are built up eg. with 8x8 lenses having a diameter of 500μm, an effieciency of 80 % and a F-number of 3 [CSEM95]. Stray light which is caused by undesired diffraction orders can couple into neighbouring channels. This is a problem especially if the lenses are used for coupling into a line or an array of fibers [Zars94].

3.2.2. Phase-gratings for spot array generation

In the case of a Fresnel lens only one focal spot is generated. But a grating can be designed to generate an array of spots. One beam of light is split into equally intense spots for applications in array illumination for eg. an array of modulators. One type of such gratings is called Dammann-grating [Krac89]. The principle is shown in fig. 3.7. Dammann-gratings are Fourier-type array generators. A parallel beam of light illuminates the grating and a lens performs a Fourier-transformation. A one-dimensional Dammann-grating can be seen in fig. 3.8. To generate equal intensity spots, one period of the grating consists of N + 1 intervals. The transmission function has only the values -1 and +1, corresponding to phase values 0 or π. The grating is completely characterized by its N transition points x_1, ... x_n where the phase changes occur. For the design shown in fig. 3.8 the number of equal intensity spots that can be generated is 2N + 1. If a regular cartesian or hexagonal spot array is desired it is sufficient to cross two one-dimensional Dammann-grating patterns.

Fig. 3.7 – A Dammann-grating acting as a beamsplitter (After [Krac89])

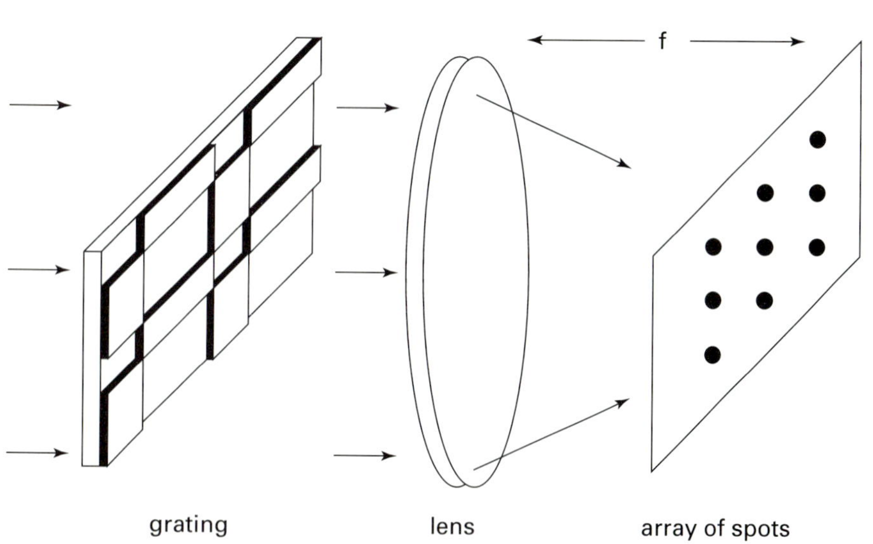

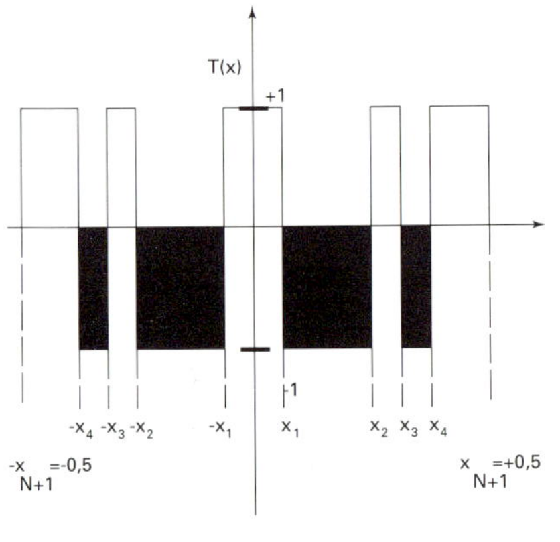

Fig. 3.8 – One period of the amplitude transmission function of a one-dimensional binary Dammann-grating

Fig. 3.9 – Partial view of the corresponding two-dimensional amplitude pattern

3.2.3. Dispersive microlenses

Based on the use of grey scale masks a fabrication method for a dispersive microlens has been reported [Gal93]. It is a composition of a blazed grating and a refractive microlens. The grating is fabricated on top of the lens as indicated in fig. 3.10. With this kind of lens simultaneously focussing and dispersing light in a wavelength band is possible. The same result is achieved if a grating can be moulded on top of a microlens (see also fig. 3.10) [Xia96].

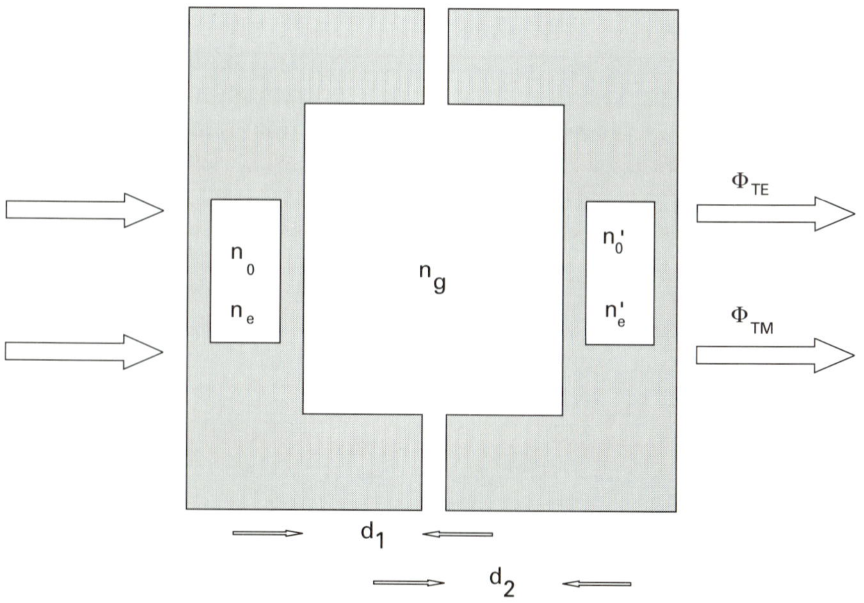

Fig. 3.10 – Dispersive microlens concept (After [Gal93, Xia96])

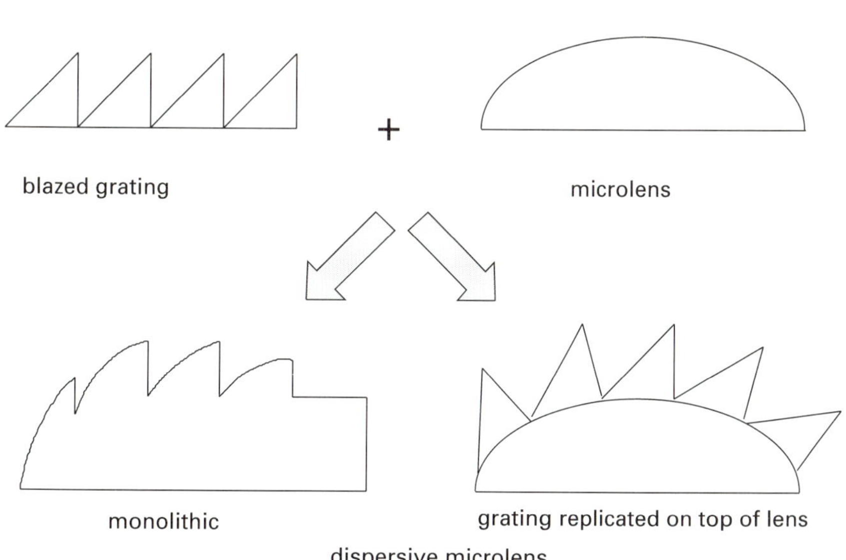

Fig. 3.11 – Schematic representation of a polarization-selective diffractive optical element (After [Nieu95])

3.2.4. Polarization-selective diffractive optics

A birefringent material in combination with two independant polarization states of light is a very elegant method to separate spatially light with different polarization states. A phase difference between the two polarization directions will result when passing through a birefringent uniaxial crystal.

If two gratings are placed opposite to each other as indicated in fig. 3.11, the phase after passing through this element for TE and TM polarized light can be given as [Nieu95]

$$\Phi_{TE} = -\frac{2\pi}{\lambda} \left[(n_0 - n_g)d_1 + (n_0' - n_g)d_2\right.$$

$$\Phi_{TM} = -\frac{2\pi}{\lambda} \left[(n_e - n_g)d_1 + (n_e' - n_g)d_2\right.$$

(3.11)

From equation 3.11 the etch depth d_1 and d_2 can be calculated

$$d_1 = \frac{2\pi}{\lambda} \frac{(n_e' - n_g)\,\Phi_{TE} - (n_0' - n_g)\,\Phi_{TM}}{(n_e - n_g)(n_0' - n_g) - (n_e' - n_g)(n_0 - n_g)}$$

$$d_2 = \frac{2\pi}{\lambda} \frac{(n_e - n_g)\,\Phi_{TE} - (n_0 - n_g)\,\Phi_{TM}}{(n_e' - n_g)(n_0 - n_g) - (n_e - n_g)(n_0' - n_g)}$$

(3.12)

The substrate of the two gratings can be the same. But when combining the two elements the optical axis must be perpendicular and the two gratings have to be carefully aligned relative to each other. The index of the gap g can be choosen arbitrarily (eg. air). A simplification is possible in the following case:

$$n_0' = n_e \, , \, n_e' = n_0 \, , \, n_g = n_e \ (\text{or } n_g = n_0)$$

Then equations 3.12 become

$$d_1 = -\frac{\lambda}{2\pi} \frac{1}{n_0 - n_e} \Phi_{TE}$$

$$d_2 = -\frac{\lambda}{2\pi} \frac{1}{n_0 - n_e} \Phi_{TM}$$

(3.13)

The etch depth for both elements is identical, $d_1 = d_2$.

Such elements have been fabricated in $LiNbO_3$ [Xu93] and in Calcite [Nieu95]. Table 3.2 gives the parameters for these two materials.

Table 3.2

material	n_0	n_e	Δn	λ/nm
$LiNbO_3$	2.3325	2.2422	0.0903	514.5
Calcite	1.66	1.49	0.17	632.8

The advantages of Calcite over $LiNbO_3$ are that Calcite is less expensive, its Δn is higher and it can be processed using wet etching. Moreover the refractive indices are closer to 1 where reflection losses are smaller.

3.3. Packaging of diffractive optics

Free space optical systems offer high parallelism and high interconnection density for data transfer. But their realization is often limited by mechanical mounting and alignment problems. High alignment precision is necessary in order to provide high coupling efficiency for detectors or optical fibers. The alignment tolerances in such systems are in the order of 1 µm. It is not impossible to achieve this accuracy but the whole assembly becomes bulky and costly. For this reason, it is of considerable interest to develop packaging techniques which guarantee the specified accuracy with respect to mechanical and thermal influences.

3.3.1 Planar Optics

In the planar optics approach the three-dimensional free space optical setup is folded into a two-dimensional geometry (fig. 3.12) [Jahn94]. The diffractive elements are fabricated by means of lithography on one or both sides of an optically flat quartz glass substrate or a semiconductor material that is transparent at the wavelength used. The light travels along a zigzag path through the substrate. Wave propagation with two - dimensional input planes is in three dimensions and therefore such a system is a true free space optical setup. Light propagation takes place in one single substrate which makes the system compact and robust against mechanical and thermal influences. The deflection angles on one hand are determined by the size of the image plane and the substrate thickness. On the other hand they are of course limited by the lithographic process itself. Typical angles range from 5° to 30°. To achieve high reflectivity at the bottom side the substrate is covered by highly reflecting coatings. Additional absorptive layers can reduce scattered light. However the inclined optical axis causes aberrations of the imaging setup which has to be compensated by a suitable element design. Several planar optical systems have already been demonstrated as for example free space optical clock distribution and highly parallel optical interconnect using 32x32 channels. Also flip chip bonding of VCSELS on a planar optical system has been reported [Jahn93, Jahn94].

Fig. 3.12 – Concept of planar optics (After [Jahn94])

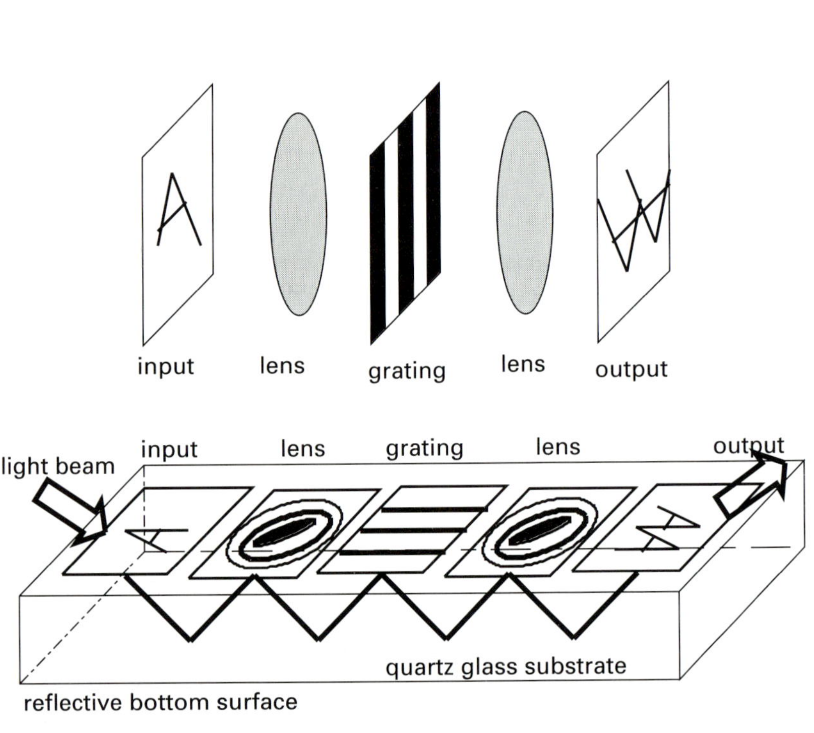

Fig. 3.13 – Planar optics with refractive field lens (After [Stre93])

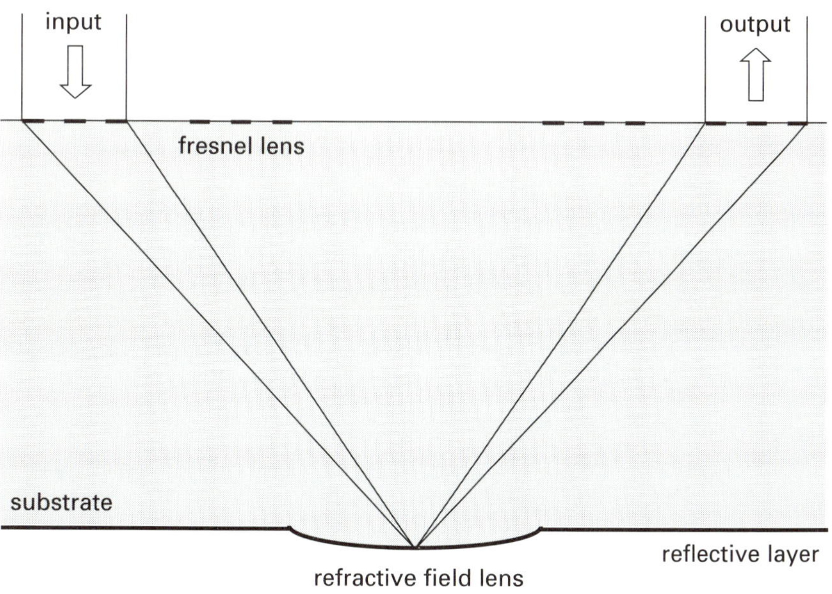

3.3.2. Planar optics combined with refractive optics

The purely diffractive approach of planar optics can be combined with refractive optical elements in order to provide better beam quality or to reduce the overall length of the setup. In fig. 3.13 a refractive field lens is added to improve the light efficiency imaged on the detector side [Stre93]. The field lens can be realized e.g. with the technique described in chapter 7. Also microprisms for data permutation can be additionally etched into a planar optics setup [Song92]. The Silicon substrate allows only angles of 70.5° for anisotropic etching. This has to be taken into account for the layout. Fig. 3.14 shows the schematic setup.

Diffractive Optics

Fig. 3.14 – Planar optics with refractive permutation prisms (After [Song92])

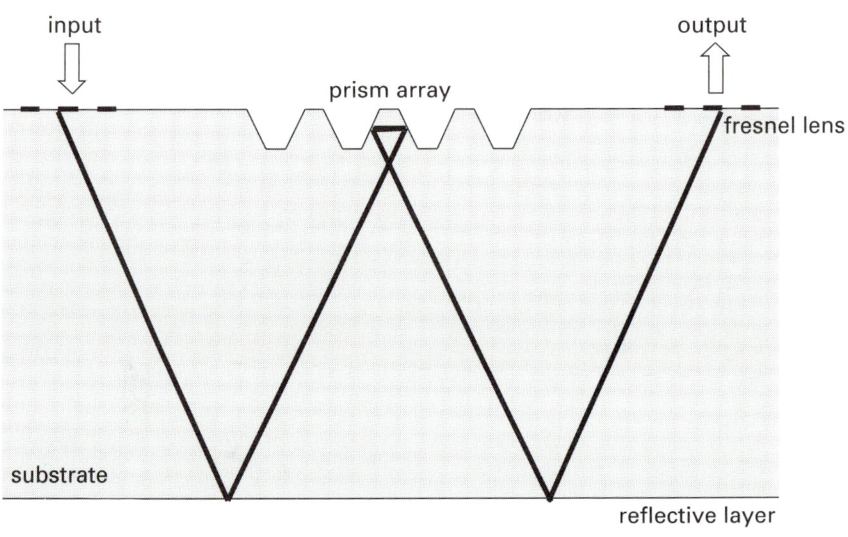

Fig. 3.15 – Schematic diagram of the three-dimensional micro-optics with side latches on both sides (After [Lin94])

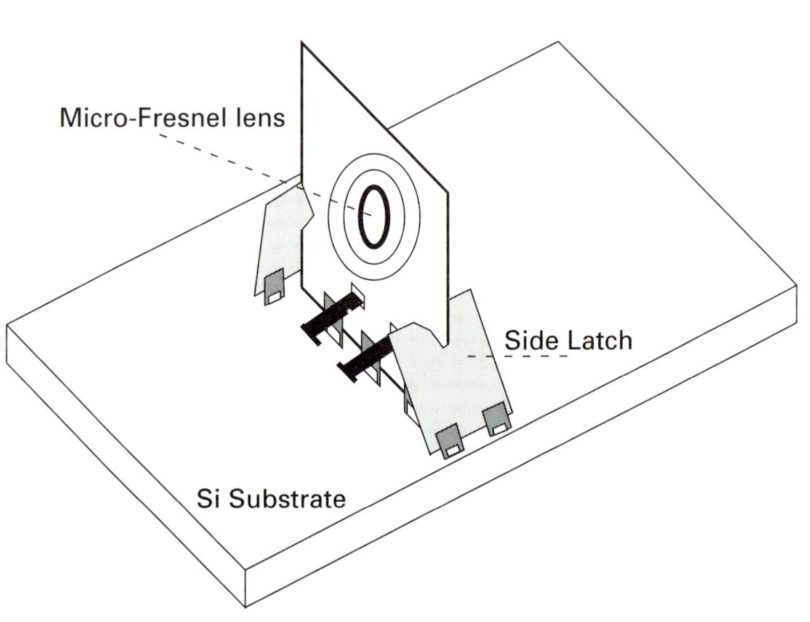

3.3.3 Surface micromaching in polysilicon

Another type of free space micro-optical integrated systems is based on micromachining on Silicon substrates. Using sacrificial layer techniques the optical elements can be perpendicular to the substrate with their bottom fixed by micro-machined micro-hinges and micro-spring latches [Lin94]. The principle is shown in fig. 3.15. Various optical elements such as Fresnel lenses, mirrors, beam splitters and gratings can be fabricated with this technique. Also a combination of passive optics with active devices like vertical-cavity surface-emitting lasers has been demonstrated succesfully [Lee95].

References for Chapter 3

[CSEM95] CSEM product information, Switzerland, 1995.

[Dhoe95] B. Dhoedt: "Theoretische en Experimentele Studie van Vrije-Ruimte Optische Interconnecties op Basis van Diffractieve Lensrijen", Thesis Uni Gent, 1995

[Gal93] G. Gal, B. Herman, W. Anderson, D. Shough, D. Purdy, and A. Berwick: "Development of the dispersive microlens", in Microlens Arrays, Vol. 2, 1993 EOS Topical Meetings Digest Series, pp. 38 - 41.

[Gale94] M. T. Gale, M. Rossi, J. Pedersen, and H: Schütz: "Fabrication of continuous-relief micro-optical elements by direct laser writing in photoresist", Optical Engineering, Vol. 33, No. 11, p. 3556 - 3566, 1994.

[Harr94] L. R. Harriott and H. Temkin in 'Integrated optoelectronics', chapter 6, "Focused ion beam fabrication techniques for optoelectronics", p. 213 - 256, ed. M. Dagenais, R. F. Leheny, and J. Crow, Academic Press, San Diego 1994.

[Jahn93] J. Jahns and B. Acklin: "Integrated planar optical imaging system with high interconnection density", Optics Letters, Vol. 19, No. 19, p. 1594 - 1596, 1993.

[Jahn94] J. Jahns in 'Optical computing hardware', chapter 6, "Diffractive optical elements for optical computers", p. 137 - 167, ed. J. Jahns and S. H. Lee, Academic Press, Boston, 1994.

[Krac89] U. Krackhardt and N. Streibl: "Design of Dammann-gratings for array generation", Optics Communications, Vol. 74, No. 1,2, p. 31-36, 1989.

[Lee95] S. S. Lee, L. Y. Lin, K. S. J. Pister, M. C. Wu, H. C. Lee and P. Grodzinski: "Passively aligned hybrid integration of 8x1 micromachined micro-fresnel lens arrays and 8x1 vericalcavity surface-emitting laser arrays for free-space optical interconnects", IEEE Photonics Technology Letters, Vol. 7, No. 9, p. 1031 - 1033, 1994.

[Lin94]   L. Y. Lin, S. S. Lee, K. S. J. Pister, and M. C. Wu: "micromachined three-dimensional micro-optics for integrated free-space optical systems", IEEE Photonics Technology Letters, Vol. 6, No. 12, p. 1445 - 1447, 1994.

[Nieu95]   N. Nieuborg, H. Thienpont, and I. Veretennicoff: " Polarisation-selective diffractive lenslet arrays", in Microlens Arrays, Vol. 5, 1995 EOS Topical Meetings Digest Series, p. 79 - 84, 1995.

[Shio94]   T. Shiono in 'Optical computing hardware', chapter 7, "Diffractive micorlenses fabricated by electron-beam lithography", p. 169 - 191, ed. J. Jahns and S. H. Lee, Academic Press, Boston, 1994.

[Song92]   S. H. Song, E. H. Lee, C. D. Carey, S. R. Selviah, and J. E. Midwinter: "Planar optical implementation of crossover interconnects", Optics Letters, Vol. 17, No. 18, p. 1253 - 1255, 1992.

[Stre93]   N. Streibl, R. Völkel, J. Schwider, P. habel, and N. Lindlein: "Parallel optoelectronic interconnectins with high packing density through a light-guiding plate using grating couplers and field lenses", Optics Communications, Vol. 99, p. 167-171, 1993.

[Sule95]   T. J. Suleski and D. C. O'Shea: "Grey scale masks for diffractive-optics fabrication: I. Commercial slide imagers", Applied Optics, Vol. 34, No. 32, p. 7507 - 7517, 1995.

[Xia96]   Y. Xia, E. Kim, X.-M. Zhao, J. A. Rogers, M. Prentiss, and G. M. Whitesides: "Complex optical surfaces formed by replica moding against elastomeric masters", Science, Vol. 273, p.347 - 349, 19 July 1996.

[Xu93]   F. Xu, J. E. Ford, and Y. Fainman: "Polarization selective computer generated holograms and applications", in Miniature and Micro-Optics and Micromechanics, N. C. Gallagher, Jr., C. Roychoudhuri, ed., Proc. SPIE 1992, p. 190 - 201, 1993.

[Zars94]   H. Zarschizky, A. Stemmer, F. Mayerhofer, G. Lefranc, W. Gramann: "Binary and multilevel diffractive lenses with submicrometer feature sizes", Optical Engineering, Vol. 33, No. 11, p. 3527 - 3536, 1994.

Guided Wave Optics - Integrated Optics

4.1. Principle of waveguiding

In 1969 the concept of integrated optics was proposed by S. E. Miller [Mill69]. It was the first optical integration method directly based on the electronic model. Lithography is used to design optical circuit layout similar to electronics. Light is guided in optical wires in materials like polymers, glass, $LiNbO_3$ or semiconductors. During more than 20 years of material development and improvement the attenuation of eg. fibers has become very low. Losses less than 0.2 dB/km have been reported [Kash95]. Waveguiding can be achieved in areas where the refractive index is higher than in the surrounding medium. Depending on whether there is a discrete step of the refractive index or a continuous transition the waveguides are classified as so-called step index and graded index devices. This chapter focusses mainly on the properties of step-index waveguides, whereas graded index components are discussed in more detail in the context of lens fabrication techniques in chapter 6.

Typical examples for step-index waveguides are shown in fig. 4.1. If wave propagation is designed for several modes (multi mode) the diameter d is in the range of 20 to 1000 μm. If only a single mode is allowed d is in the range of 3 to 10 μm. The acceptance angle for light to be coupled into the waveguide and the number of modes that can travel in a waveguide are functions of the refractive indices, the wavelength and the guide thickness. To explain some of the basic principles of guided wave optics a simple ray optic model as shown in fig. 4.2 can be used [Gunt90]. First the maximum acceptance angle and the numerical aperture are calculated and then the number of allowed modes given by the cut-off condition. For simplicity only TE mode theory is calculated. It is also assumed that $n_2 > n_3 > n_1$.

Fig. 4.1 – Examples of waveguides

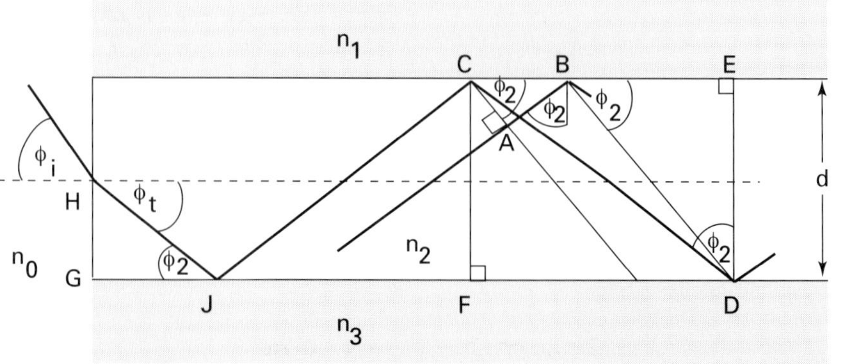

The angle ϕ_2 in fig. 4.2 must exceed the critical angle of total internal reflection (TIR).

$$\Phi_2 \geq \Phi_c = \sin^{-1} \frac{n_3}{n_2}$$

ϕ_2 can also be expressed in terms of the triangle JGH. It follows

$$\left(\frac{GH}{JH}\right)^2 = 1 - \left(\frac{JG}{JH}\right)^2 \leq 1 - \left(\frac{n_3}{n_2}\right)^2$$

Fig. 4.2 – Ray optic model for guided waves

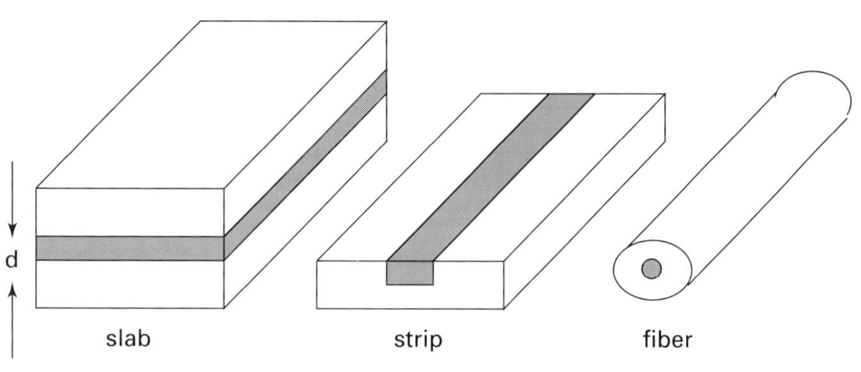

The sides of the triangle can also be expressed using ϕ_t:

$$\frac{GH}{JH} = \sqrt{1-\left(\frac{n_3}{n_2}\right)^2} = \sin\Phi_t$$

Snell's law relates the transmission angle ϕ_t to the angle of incidence ϕ_i on the fiber face:

$$\frac{n_2}{n_0} = \frac{\sin\Phi_i}{\sin\Phi_t}$$

The acceptance angle can than be written in terms of the above relations:

$$\sin\Phi_i \leq \sin\Phi_{NA} = \frac{n_2}{n_0}\sqrt{1-\left(\frac{n_3}{n_2}\right)^2}$$

The numerical aperture NA of the guide can now be defined as

$$NA = n_0\sin\Phi_{NA} = n_2\sin\Phi_t$$
$$= \sqrt{n_2^2 - n_3^2} \tag{4.1}$$

The result of equation 4.1 shows that the numerical aperture is independant of the core radius. And it can be seen that with a higher difference in the refractive indices the numerical aperture increases.

There are also restrictions to the number of modes; this can be seen in fig. 4.2 following a plane wave propagating through the waveguide of index n_2 and thickness d: Rays at the wavefront BD have travelled different optical path lengths, e.g. rays CD and AB have to be compared. The phase difference is given by equation 4.2 and must be a multiple of 2π.

$$n_2k\,(CD - AB) + \delta 1 + \delta 3 = m2\pi \tag{4.2}$$

δ_1 and δ_3 are additional phase shifts that have to be taken into account because of the Goos-Hähnchen-Shift which occurs at total internal reflection. Using the trigonometrical relationships given by fig. 4.2 equation 4.2 can further be transformed:

$$2n_2kd\cos\Phi_2 + \delta 1 + \delta 3 = m2\pi \tag{4.3}$$

It can be seen that ϕ_2 can have only discrete values dependant on m. In addition ϕ_2 is also limited to

$$\sin^{-1}\frac{n_3}{n_2} < \Phi_2 < \frac{\pi}{2} \tag{4.4}$$

Because of the relationship 4.4 which is given by Snell's law it follows

$$\cos\Phi_m < \sqrt{1-\left(\frac{n_3}{n_2}\right)^2} \tag{4.5}$$

Equation 4.3 can be rewritten using equation 4.5:

$$m_{max} \le \frac{2dn_2}{\lambda_0}\sqrt{1-\left(\frac{n_3}{n_2}\right)^2} + \frac{\delta1+\delta3}{2\pi} \tag{4.6}$$

The quantity V is defined as

$$V = kdn_2\sqrt{1-\left(\frac{n_3}{n_2}\right)^2}$$
$$= kd\sqrt{n_2^2 - n_3^2} \tag{4.7}$$
$$= kd\ NA$$

V is called the normalized film thickness for the case of planar waveguides and normalized frequency (or V number) for optical fibers.

Using the definition of V equation 4.6 can be written as

$$m_{max} \le \frac{V}{\pi} + \frac{\delta1+\delta3}{2\pi} \tag{4.8}$$

A first estimation for m_{max} can be given assuming that the phase shift upon reflection is smaller than 180°.

$$m_{max} \le \frac{V}{\pi} + 1 \tag{4.9}$$

Guided Wave Optics – Integrated Optics

But more precisely δ_1 and δ_3 are given by [Gunt90]

$$\delta_1 = -2\,\tan^{-1}\frac{\sqrt{\sin^2\Phi_2 - (\frac{n_1}{n_2})^2}}{\cos\Phi_2} \tag{4.10}$$

$$\delta_3 = -2\,\tan^{-1}\frac{\sqrt{\sin^2\Phi_2 - (\frac{n_3}{n_2})^2}}{\cos\Phi_2}$$

In the literature equation 4.3 is often defined by other variables which are called the effective index N, the normalized index b and the asymmetry parameter a [Nish89].

$$N = n_2\,\sin\Phi_2 \tag{4.11}$$

$$b = \frac{N^2 - n_3^2}{n_2^2 - n_3^2} \tag{4.12}$$

$$a = \frac{n_3^2 - n_1^2}{n_2^2 - n_3^2} \tag{4.13}$$

From equation 4.4 it follows that $n_3 < N < n_2$, $0 < b < 1$ and $a > 0$ (if $n_2 > n_3 > n_1$). Using these new definitions equation 4.10 can be written as

$$\delta_1 = -2\,\tan^{-1}\sqrt{\frac{b+a}{1-b}}$$

$$\delta_3 = -2\,\tan^{-1}\sqrt{\frac{b}{1-b}} \tag{4.14}$$

For equation 4.3 it follows

$$2k\,d\,n_2\,\cos\Phi_2 - 2\,\tan^{-1}\sqrt{\frac{b+a}{1-b}} - -2\,\tan^{-1}\sqrt{\frac{b}{1-b}} = m\,\pi \tag{4.15}$$

Or using the V parameter (equation 4.7)

$$2V\sqrt{1-b} - 2\,\tan^{-1}\sqrt{\frac{b+a}{1-b}} - -2\,\tan^{-1}\sqrt{\frac{b}{1-b}} = m\,\pi \tag{4.16}$$

A few examples will demonstrate the use of relationship 4.16:

When ϕ_2 is equal to the critical angle of TIR ϕ_c it follows

$$\sin\Phi_2 = \frac{n_3}{n_2}$$

and for the effective index N

$$N = n_2 \sin\Phi_2 = n_3$$

Hence b = 0. Relationship 4.16 becomes

$$V = \tan^{-1}\sqrt{a} + m\,\pi \qquad (4.17)$$

For a symmetric guide ($n_1 = n_3$) the asymmetry parameter is zero, a = 0 and therefore $V = m\pi$. If m = 0 it can be seen from equation 4.7 that there are no restrictions (guide thickness, wavelength) for the zero mode to propagate in a symmetric guide. If m = 1, than $V = \pi$. The guide thickness can be calculated to be

$$d = \frac{\lambda_0}{2\sqrt{n_2^2 - n_3^2}}$$

If λ_0, n_2, and n_3 are given, than d represents the smallest guide thickness that will allow mode 1 to propagate (together with mode 0). If the guide thickness is smaller, the guide is said to be single mode which means that only mode 0 can propagate. In all other cases the guide is said to be multimode.

For an asymmetric guide ($n_1 \neq n_3$), it follows a > 0. This means that mode 0 will not propagate if the guide thickness is less than a certain minimum given by equation 4.7 and 4.17. The number of modes in a multimode guide is m+1, where m is the largest mode that can propagate.

Fig. 4.3 – Waveguide configurations: step index or graded index waveguides

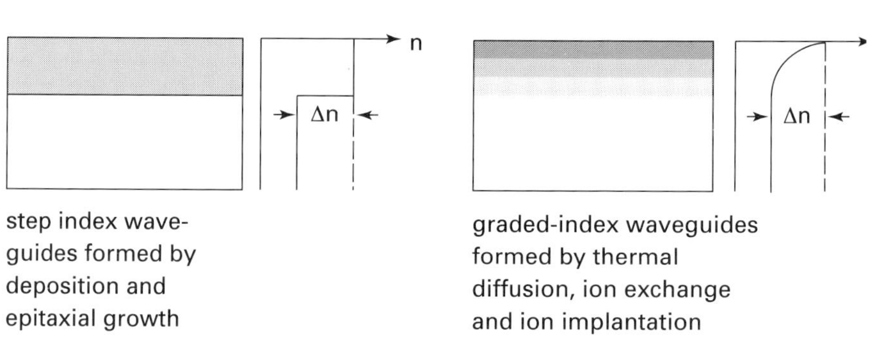

4.2. Materials and fabrication techniques for optical waveguides

Single mode waveguides in integrated optics are especially attractive if switching control has to be applied. Layed out in materials such as glass, polymers and particularly $LiNbO_3$ electrooptic, acoustooptic or thermooptic effects are possible. The material must have a high degree of optical purity in order to provide low attenuation (< 1 dB/cm). The lithographic process on one substrate assures that no further alignment is needed and that the arrangement is stable with respect to vibrations and temperature. The setups are compact and very light weight.

The waveguides can be fabricated in different ways, as step index or graded index guide (fig. 4.3). In the following each fabrication technique will be briefly explained.

Thermal Vapor Deposition
Thin films are deposited onto a substrate. The vaporization occurs in a vacuum chamber by heating. A film thickness of more than 2 μm with good homogeniety and adhesion is difficult to fabricate.

Sputtering
By discharge positive ions collide with a target plate. Particles of the target plate are knocked out and deposited on a substrate. Deposition rates can be as high as 5 μm/min depending on the sputter technique. Films are very pure, strong and in general more uniform than thermally vaporized films.

Epitaxial Growth
Two crystal substrates with similar lattice constants can be grown together. Liquid phase epitaxy and vapor phase expitaxy exist.

Thermal Diffusion
A material to be diffused is deposited onto a substrate and structered lithographically. Then the substrate is put into an oven for a certain time where the deposited material starts to diffuse into the substrate. This method is used eg. for the indiffusion of Ti into $LiNbO_3$.

Ion Implantation
Ion implantation can result in high refractive index change. Ion depth can be controlled via the ion energy and the ion current determines the change in refractive index. This method can be used for eg. implanting B^+ ions into fused-silica substrates or H^+ ions into plastics. Examples of plastic waveguides fabricated with this technique will be discussed in chapter 10.

Thermal Ion Exchange
This technique is used especially for glass substrates and $LiNbO_3$. A slight index change takes place where ions such as Ag^+, K^+, Tl^+ can diffuse into the substrate and replace lighter ions like Na^+. Typical diffusion times are in the order of a few hours. The process can be accelerated if an electrical field is applied along the diffusion direction of the ions. Then the mobility and penetration depth of the ions is much higher resulting in a reduced diffusion time but also in a different index profile. (This process is described in more detail in chapter 6.)

Spin Coating
This is the classical lithographic process using polymers on a substrate (eg. glass). After structuring and development the resist represents the waveguide layout. This is the simplest and least expensive approach but purity and uniformity of the films is rather poor.

Buried polymeric waveguides
A simple modification of this process is much better suited for waveguide fabrication [Neye94]. The photoresist is not used for guiding but only as auxilliary structure. The fabrication process is shown in fig. 4.4.

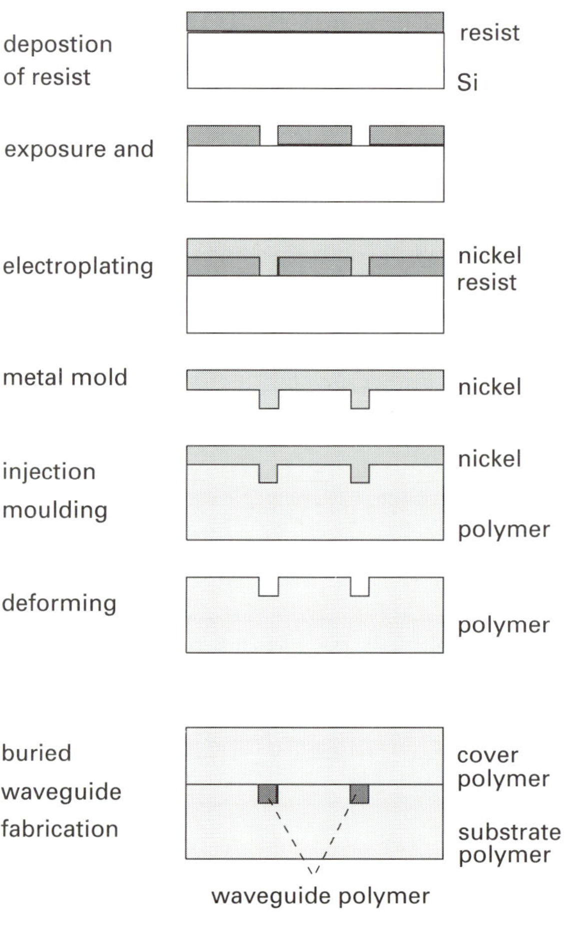

Fig. 4.4 – Schematic diagram of fabrication of buried polymeric waveguides using injection moulded substrates (After [Neye94])

A photoresist layer is spun onto the substrate. The thickness of the resist defines the final waveguide thickness. With UV exposure and development waveguide grooves are etched in the resist. Electroplating in nickel yields a mould insert. With injection moulding many copies of the former master can be made eg. in PMMA [Bara93]. The precision of the electroplating is in the submicrometer range. The last step in fabrication is to fill the grooves with deuterated EGDMA (ethylenglycol dimethacrylate) that shows less than 1 dB/cm insertion loss at 1300 nm. After filling, a plane plate of PMMA is pressed against the substrate to remove the surplus liquid material and to protect the waveguide from the top. An intermediate layer of oligomer with a thickness of about 1 µm remaines between the two PMMA substrates. The last step is to cure the oligomere in the grooves by UV light. This technique has a potential for mass fabrication. It can also provide alignment pedestals and stand-offs. A variety of materials can be choosen as cores to provide passive or optically nonlinear guides.

Flexible waveguides
V-Grooves are necessary to hold fibers which have a large diameter compared to the waveguides. Optical fibers are required to provide a connection to the outside world due to the relative inflexibility of the waveguide. However this is not required to flexible waveguides which have no substrate. This has been shown in [DeDo95]. There a technique has been developed for the fabrication of flexible single mode wave-guides. The layer structure is shown in fig. 4.5. If the substrate and its layers on top are brought in a hot KOH bath, the gold layer is etched away and the three layer waveguide is stripped from the substrate. It is a very flexible and reliable guide that can easily be fabricated in a bundle containing a line of waveguides. A photograph of such a flexible line of waveguides can be seen in fig. 4.6.

Fig. 4.5 – Substrate composition for the fabrication of flexible waveguides (After [DeDo95])

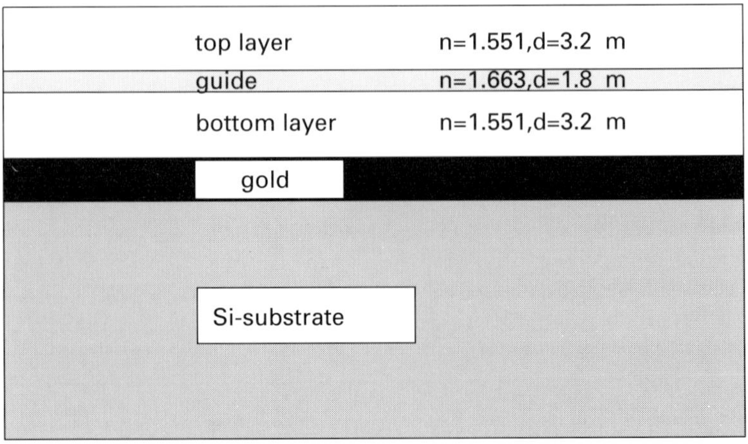

top layer	n=1.551,d=3.2 m
guide	n=1.663,d=1.8 m
bottom layer	n=1.551,d=3.2 m
gold	
Si-substrate	

Fig. 4.6 – Photograph of a flexible waveguide bundle (courtesy Peter De Dobbelaere [DeDo95])

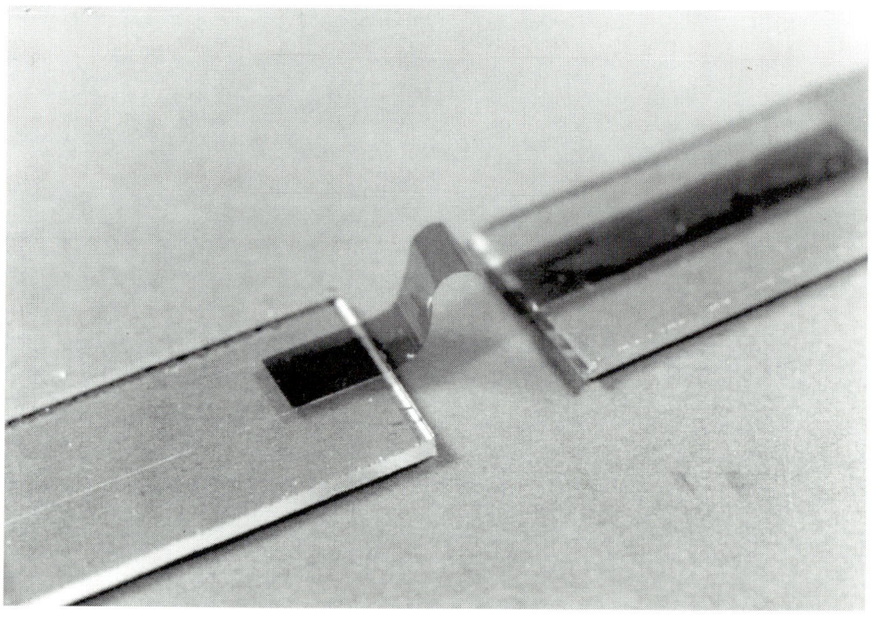

4.3. Alignment of waveguides

Inside a waveguide network no further alignment is required. This is one of the major advantages of this technique. But for connection to the outer world incoupling and outcoupling circuits must be adopted. Several solutions for fiber attachement to waveguides have been reported in the literature, depending on the choice of the waveguide substrate. Drawbacks are that plastic waveguide substrates cannot withstand the high temperature treatment which is necessary for flip-chip bonding. And in $LiNbO_3$ no V-grooves can be etched directly. Only a U-groove is possible whose alignment tolerances are however not as good as with V-grooves.

V-groove fabrication
Very often the alignment is done by so called V-grooves in the case of fibers. Special interest is given to single mode circuits where the alignment tolerance is of the order of ± 1 µm. In the case of a (100)-oriented

silicon substrate there exists an etch anisotropy between the (100) and (111) planes of the order of 400 : 1. With alcaline solutions such as NaOH etch angles of 54,74° can be achieved (a similar technique has already been discussed for the fabrication of microprisms in chapter 3.4.2). This technique can also be applied for multilayers as indicated in fig. 4.7. [Mill78].

Fig. 4.7 – V-grooves in (100) oriented Silicon

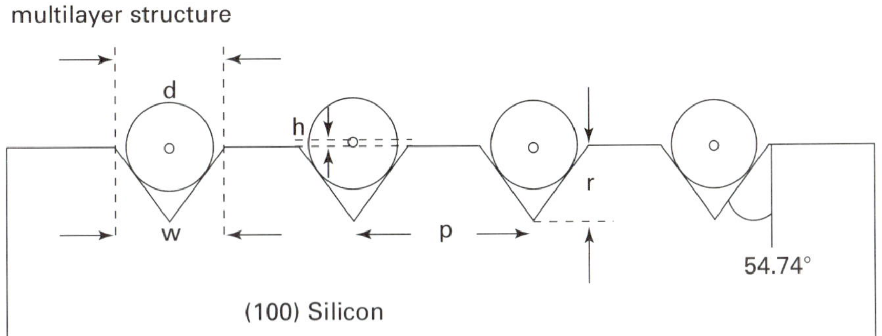

In fig. 4.7. the abreviations are

d	fiber diameter
w	etch width
p	pitch
r	etch depth
h	height of core center (relative to substrate surface)

The V-groove technique is more than 25 years old but still under investigation. The lateral position of a groove (pitch) is controllable with high accuracy of better than 1 μm. More difficult is the etch depth r. Underetching occurs during the etch process which limits the etch depth accuracy to about ± 1 μm. Additionally the diameters of fibers show tolerances in the order of about ± 1 µm. For a multilayer system as in fig. 4.7. this makes a singlemode layout very difficult because there can be a lateral displacement of the fiber from the center of the groove. Normally the fiber touches the two tilted walls of the groove. In case of a multilayer structure with lateral misalignment the fiber will touch only one side of each V-groove (fig. 4.7). This displaces a fiber in x and y direction. A spring system in silicon has been proposed to overcome this problem but it seems to be practicable only for a single fiber [John93]. To give the fibers higher stability in the V-grooves asymmetric fibers can be drawn that have the same cross-sectional profile as the V-groove [Pigr94]. Meanwhile V-grooves are also commercially available products with a specified waferscale accuracy of 0.6 µm [CSEM95].

Flip-chip techniques
Flip-chip bonding is a self aligning technique that makes use of the reflow of molten solder bumps. A schematic is shown in fig. 4.8. Each of the two substrates has small pedestals that must be aligned relative to each other. On one substrate the pedestals are covered by eg. an indium solder. With vacuum picks the two substrates are roughly aligned over each other and heated. When the solder melts it balls up and contacts the upper substrate at the position of the pedestals. During cooling the reflow of the solder brings the two substrates in the right position and fixes them. Alignment tolerances vertically and horizontally are in the order of ± 1 µm. The solder melting point for a Pb/Sn alloy (95%/5%) is given to be 350°C. The so called C^4-technology can align with this technique waveguides on V-groove substrates [Wale90, Ackl95]. Flip-chip bonding is also in use for laser diode arrays and on V-grooves [Armi91, Hunz95].

Fig. 4.8 – Schematic of flip-chip bonding

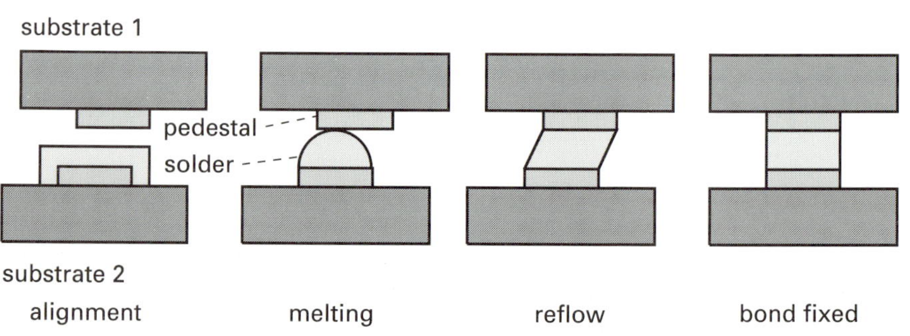

Fig. 4.9 – Schematic of the index method (After [Cohe91])

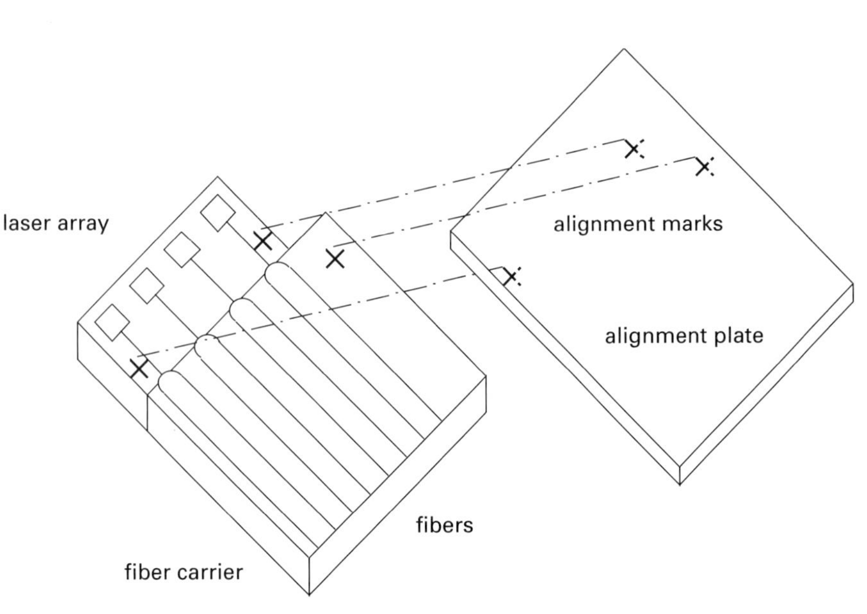

Index method
The socalled index method needs a third substrate for the alignment of waveguides and V-grooves or laser chips and V-grooves [Cohe91]. Marks on the both substrates have to be aligned relatively to those on the third substrate as shown in fig. 4.9. After alignment flip-chip bonding holds all three substrates together.

Active alignment
The previous methods use only passive alignment. Active alignment also exists. In that case fixing of different substrates depends on measured signal output during alignment. But from an economical view point passive alignment is preferred.

Other alignment techniques will be discussed in chapter 6, 8, 9 and 10 using GRIN-lens glass sockets, excimer ablation, LIGA and deep proton lithography.

Fig. 4.10 – Basic geometries of waveguides (After [Sale91])

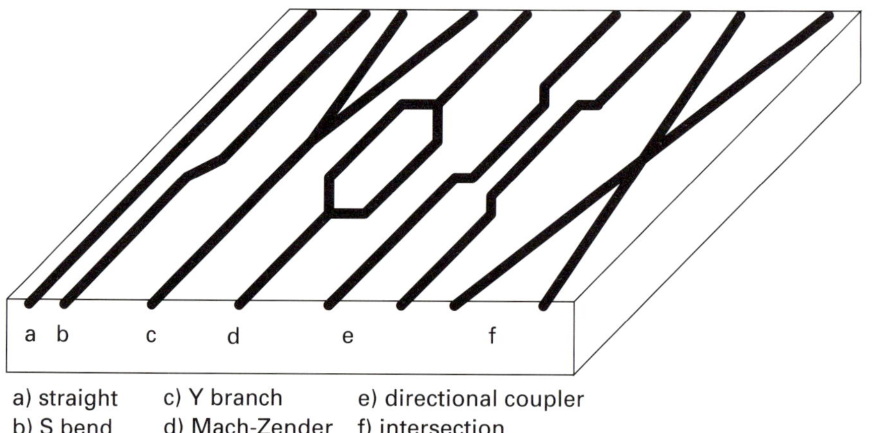

a) straight   c) Y branch   e) directional coupler
b) S bend   d) Mach-Zender   f) intersection

4.4. Waveguide applications

Waveguides can be set up in various geometries as illustrated in fig. 4.10 [Sale91]. The simplest case is a straight connection (a). S-bends (b) are used to offset the optical axis. Y-branches (c) can act as a beamsplitters or as directional switches. The Mach-Zender interferometer(d) consists of two Y branches. It is mainly used in sensor applications. In a directional coupler (e) the amount of light going from one branch into the other is a function of the coupling length. Directional coupling can also be achieved by an intersection (f). Special attention has been given to integrated optics because of the switching option. Each of the shown geometries can be fabricated in addition with a switch that controlles the mode propagation. For active applications the material of choice is $LiNbO_3$ but also new kind of polymers which are currently under investigation.

Fig. 4.11 – Y branch with Al-electrodes:a) device structure, b) enlarged view of branching point, c) guided mode propagation (After [Nish89])

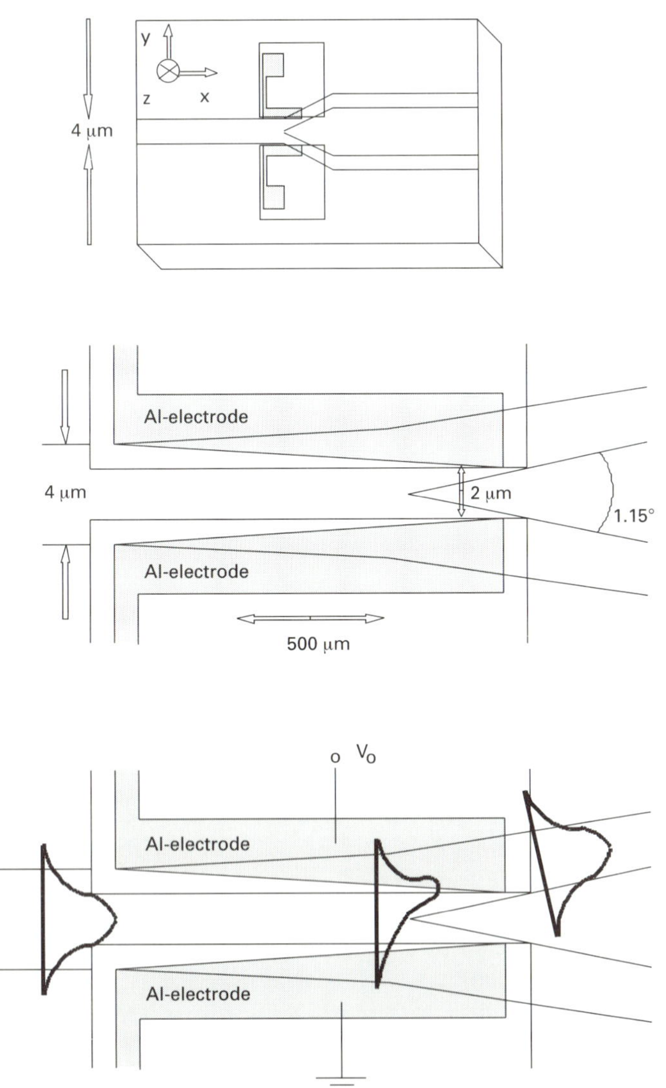

Fig. 4.11. shows a typical layout of a Y branch with given dimensions where the applied voltage determines the direction of the mode propagation [Nish89]. The dimensions are designed for singlemode waveguides. The substrate is Z-cut $LiNbO_3$ where the electrooptical effect can be used. However the device needs a certain interaction length for the switching.

In fig. 4.12. a combination of Y branches is used to build a 1 to N splitter [Hunt94]. The input can be switched to any output. In general, N-1 switches in ld N stages are required. This setup in reverse acts as an active combiner.

Another type of optical element is the directional coupler [Nish89]. Fig. 4.13. shows 3 different types. The coupling length L_0 and the gap g between the guides define the amount of light that couples from one waveguide into the other. In type a) all the light goes into the neighbouring waveguide. Type b) represents a so-called 3 dB coupler because the power is split into two equal parts. Type c) is a directional coupler where the coupling efficiency is controlled via the applied electrical voltage. The gap g in fig. 4.13 is in the order of a few microns. The coupling length which also depends on the design wavelength is in the order of several millimeters.

Fig. 4.12 – 1 to 8 splitter with 7 switching nodes in 3 stages
(After [Hunt94])

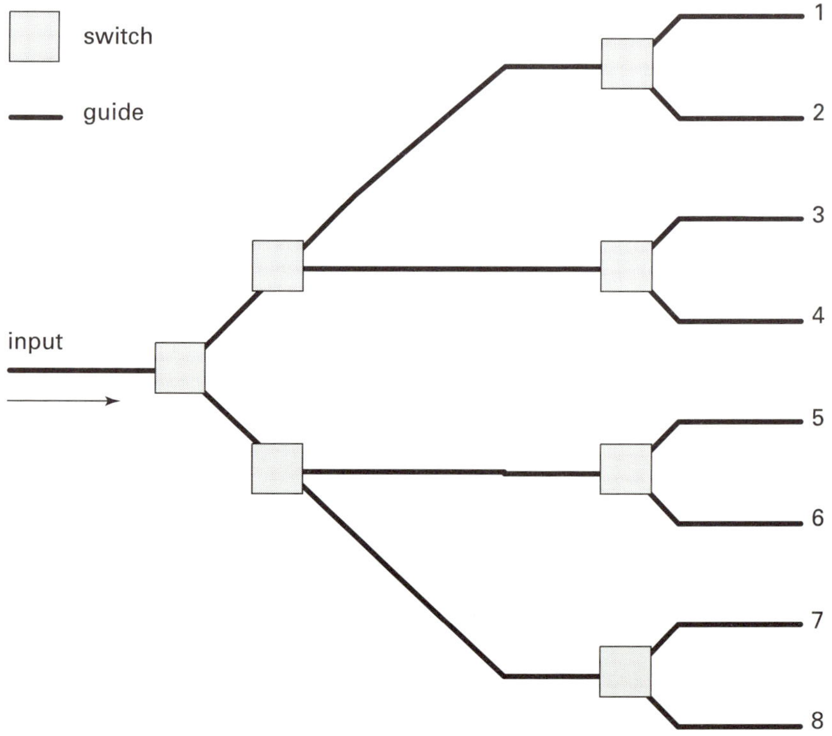

Fig. 4.13 – Different types of directinal couplers: a) all light is coupled into the neighbouring guide, b) 3 dB coupler, c) active coupler
(After [Nish89])

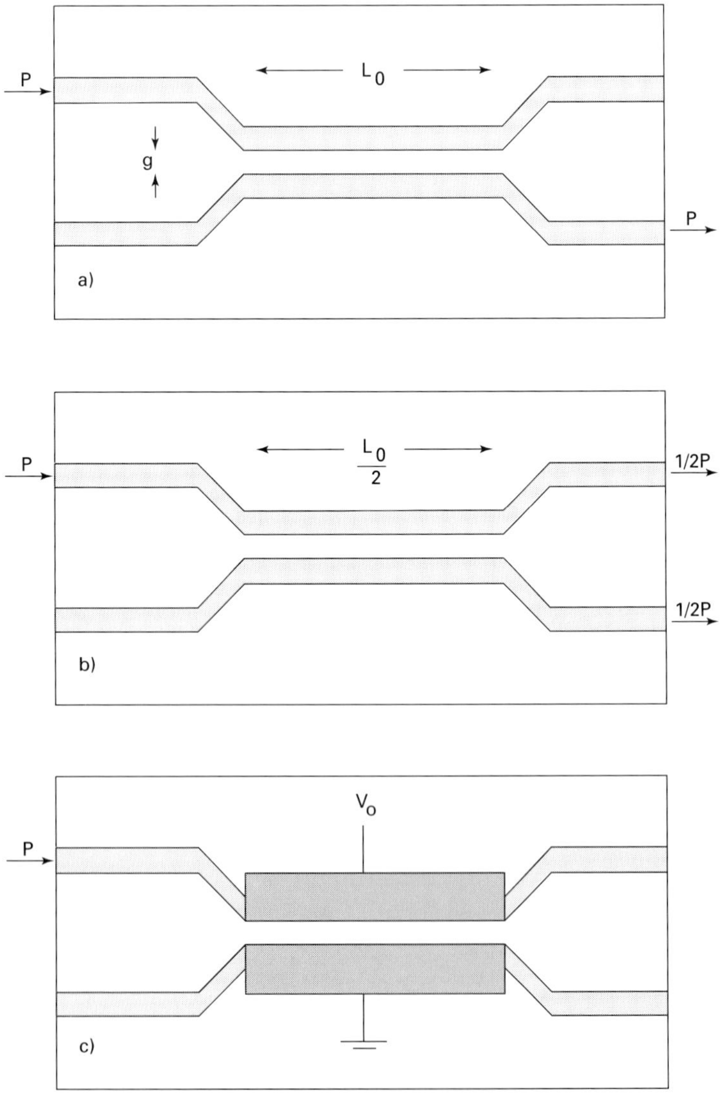

Fig. 4.14 – 4x4 optical switch network using directional couplers
(After [Nish89])

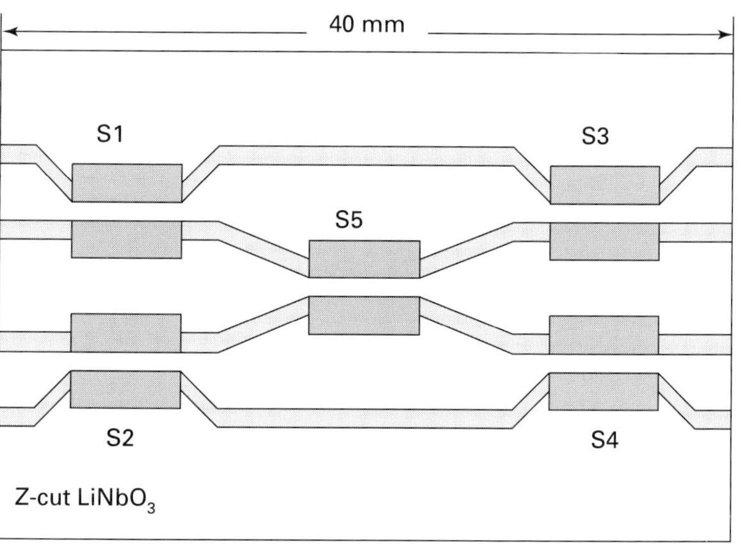

Fig. 4.14 shows a 4x4 optical switch network [Nish89]. The substrate used is Z-cut LiNbO$_3$. The waveguides are 10 µm wide with a gap of 4 µm. The coupling length is choosen to be 8 mm for a wavelength of 1.3 µm. The switches allow a bypass/exchange operation. The network length is 40 mm. If a network with 8x8 switches or more is designed the overall length is more than 8 cm. These kind of networks are therefore limited by the available wafer size. Even larger networks are fabricated on two separated substrates. Sophisticated alignment procedures are required for a stable and robust setup [Hunt94].

References for chapter 4

[Ackl95]   B. Acklin, J. Bellermann, M. Schienle, L. Stoll, M. Honsberg, and G. Müller: "Self-aligned packaging of an optical switch array with integrated tapers", IEEE Photonics Technology Letters, Vol. 7, No. 4, p. 406 - 408, 1995.

[Armi91]   C. A. Armiento, M. Tabasky, C. Jagannath, T. W. Fitzgerald, C. L. Shieh, V. Barry, M. Rothman, A. Negri, P. O. Haugsjaa, and H. F. Lockwood: "", Electron. Lett. 27, p. 1109 - 1111, 1991.

[Bara93]   L. Baraldi, R. E. Kunz, and J. Meissner: "High-precision molding of integrated optical structures", in Miniature and Micro-Optics and Micromechanics, N. C. Gallagher, Jr., C. Roychoudhuri, ed., Proc. SPIE 1992, p. 21 - 28, 1993.

[Coh91]   M. S. Cohen, M. F. Cina, E. Bassous, M. M. Oprysko, and J. L. Speidell: "Passive laser-fiber alignment by index method", IEEE Photon. Tech. Lett. vol 3, no. 11, p. 985 - 987, 1991.

[CSEM95]   CSEM product information on Micro-Systems, Neuchâtel, Switzerland, 1995.

[DeDo95]   P. De Dobbelaere: "Technologische realisatie van polymere golfgeleiderstructuren en integratie ervan met III-V-componenten en optische veszels", Thesis University of Gent, 1995.

[Gunt90]   R. Gunther: "Modern optics", Wiley, New York 1990.

[Hunt94]   D. K. Hunter and I. Andonovich: "Guided wave optical swith architectures, part 1, space switching", Int. Journal of optoelectronics, vol. 9, no. 6, p. 477 - 487, 1994.

[John93]   M. J. Johnson: "Precise alignment of optical fibers using a microfabricated silicon device", in Miniature and Micro-Optics and Micromechanics, N. C. Gallagher, Jr., C. Roychoudhuri, ed., Proc. SPIE 1992, p. 14 - 20, 1993.

[Kash95]   N. Kashima: "Passive optical components for optical fiber transmission", Artech House, Boston 1995.

[Mill69]   S. E. Miller: "Integrated optics: an introduction", Bell Syst. Techn. J. 48 (7), 2059 - 2068, 1969.

[Mill78]   C. M. Miller: "Fiber-optic array splicing with etched silicon chips", Bell System Technical Journal, Vol. 57, No. 1, p. 75 - 97, 1978.

[Neye93]   A. Neyer, T. Knoche, and L. Müller: "Fabrication of low loss polymer waveguides using injection moulding technology", Electronics Letters, Vol. 29, No. 4, p. 399 -401, 1993

[Nish89]   H. Nishihara, M. Haruna, and T. Suhara: "Optical integrated circuits", McGraw Hill, New York 1989.

[Pigr94]   P. J. Pigram, P. Y. Timbrell, R. N. Lamb, M. G. Sceats, and A. L. Cox: "Keyed optical V-fiber to silicon V-groove interconnects", Optical Engineering, Vol. 33, No. 8, p. 2594 - 2599, 1994.

[Sale91]   B. E. A. Saleh and M. C. Teich: "Fundamentals of Photonics", Wiley, New York 1991.

[Wale90]   M. J. Wale and C. Edge: "Self-aligned flip-chip assembly of photonic devices with electrical and optical connections", IEEE Transactions on Components, Hybrids, and Manufacturing Technology, Vol. 13, No. 4, p. 780 - 786, 1990.

Micro-optics in Photosensitive Glass

For several years glass types have been fabricated which are sensitive to UV light irradiation. After UV exposure the glass changes its chemical and physical properties. The interesting physical effect for micro-optics is densification of the irradiated regions. An important chemical effect is the increase in etch anisotropy of the irradiated regions.

5.1. Fabrication process

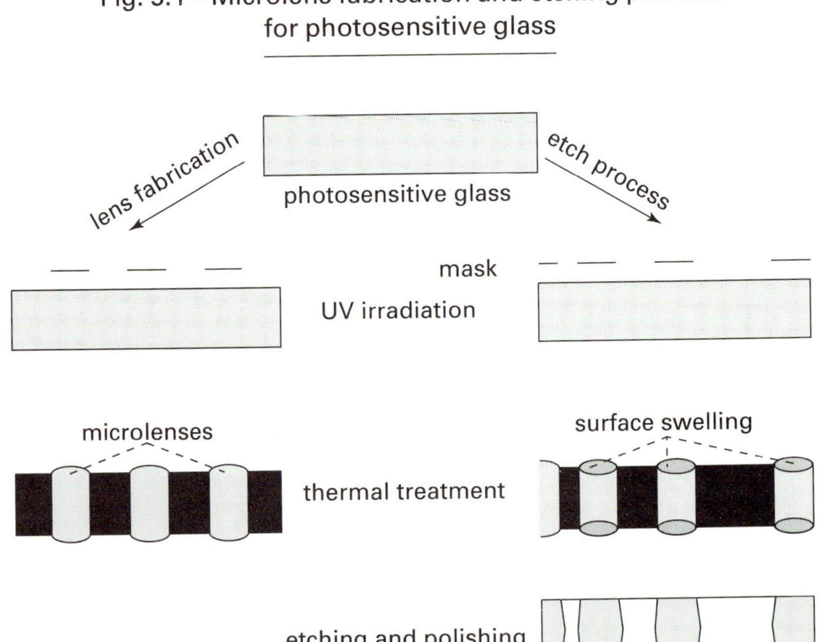

Fig. 5.1 – Microlens fabrication and etching process for photosensitive glass

Fig. 5.1 shows two different processes using either the physical or the chemical changes. The left branch describes microlens fabrication, the right branch a deep etch process in glass.

5.1.1. Microlens fabrication

After UV exposure the glass substrate undergoes a thermal treatment with temperatures of about 600°C which is above the softening temperature of the nonirradiated virgin glass [Borr85]. During the thermal cycle the irradiated regions tend to crystallize and densify. The exposed glass contracts in three dimensions and forces the unexposed glass to erupt. If the lateral dimensions are less than 1 mm the eruption causes sperical shapes or microlenses. In addition the thermal process renders the area surrounding the lens optically opaque which provides in addition an aperture for the lenses.

The shrinkage changes the overall pitch and the diameter of the lenses as indicated in fig. 5.2. and has to be taken into account during the design process.

Fig. 5.2 – Schematic representation of photothermal development of a microlens (After [Borr85])

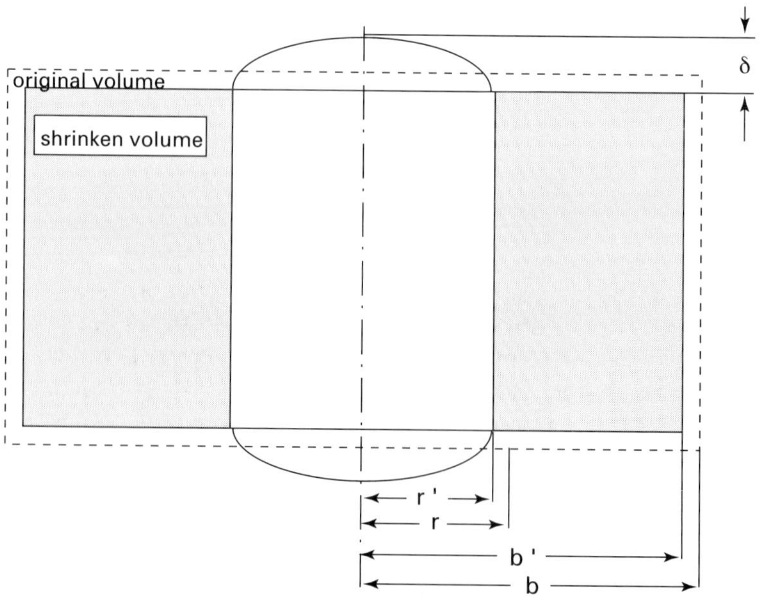

Micro-optics in Photosensitive Glass

The sag of the lens can be expressed as

$$\frac{\delta}{T} = \frac{2}{3} \left(1 - \frac{\rho}{\rho_0}\right)$$

where
  δ sag of lens
  T glass thickness
  ρ density of unexposed glass
  ρ_0 density of crystallized glass

Fig. 5.3 – Sag as a function of exposure level (After [Borr85])

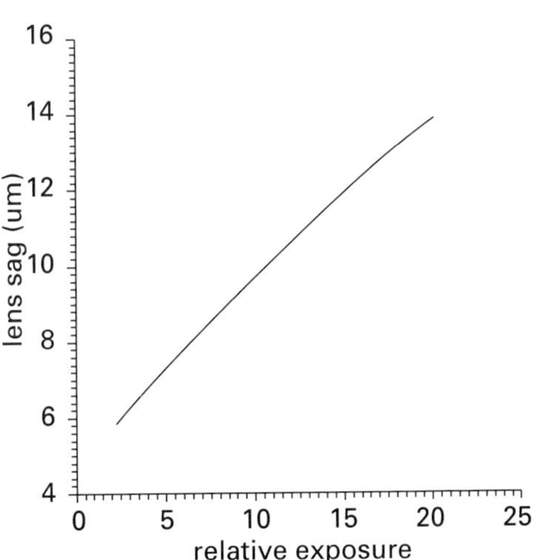

The density of the exposed regions is a function of exposure and of the thermal cycle. Fig. 5.3 shows how the sag varies with exposure. Typical diameters of lenses are in the range between 75 and 1000 µm. Lenses may not be to close to each other with a minimum seperation of 15 µm. Numerical apertures NA <= 0.35 have been reported [Borr85, Borr88]. A typical photo of such lenses can be seen in fig. 5.4. Note that only the lenses are transparent.

Fig. 5.4 – Photolytic lenses with diameters of about 160 mm
(test sample courtesy Corning Inc.)

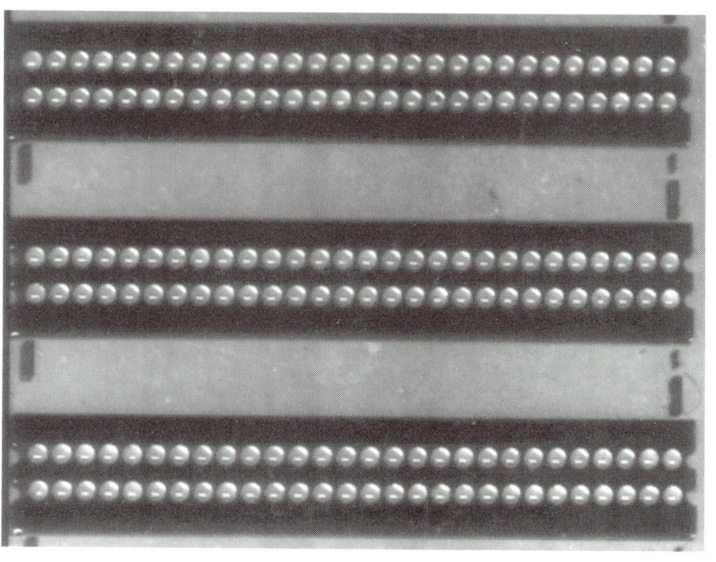

5.1.2. Anisotropic etching

Furthermore the crystallized regions show high etch anisotropy. The etch process is shown schematically in fig. 5.1. The etch attacks the irradiated regions a factor 20 to 40 faster than the nonirradiated regions. Nevertheless the nonirradiated substrate is also etched. This gives a limit for the minimum feature size of this process. Table 5.1 lists the tolerances for etched holes [Scho88].

Table 5.1

thickness/mm	smallest Ø/µm	tolerances/µm	smallest pitch/µm
0.2-0.4	60	±15	160
0.5-1.0	120	±20	250
1.1-1.5	180	±30	350
1.6-2.0	400	±40	700

An improved process from Corning promises better hole size tolerances of only ±7 µm [Corn89]. The maximum ratio of feature size to thickness is given as 40:1. The side wall angle is 1 to 2°. The minimum substrate thickness is 100 µm.

If necessary the etched glass substrate can be transferred into a ceramic. First global UV irradiation again is applied followed by a new heat treatment. Higher temperatures lead to ceramics with different properties. A typical glass composition is as follows (in wt. %)

SiO_2 75 - 85 %, Li_2O 7 - 11 %, Na_2O 1 - 2 %, K_2O 3 - 6 %, Al_2O_3 3 - 6 %, ZnO 0 - 2 %, Ce_2O_3 0.015 %, Sb_2O_3 0.3 %, Ag 0.1 %.

Fig. 5.5 shows a hexagonal hole pattern achieved with the etch process in a 1 mm thick substrate.

Fig. 5.5 – Hexagonal structures in a substrate of 1 mm thickness
(test sample courtesy of Schott Glaswerke)

5.2. Applications of photosensitive glass

5.2.1. Applications of microlenses

Photocopiers

Microlenses fabricated by the described technique have found applications in eg. photocopiers replacing gradient-index lenses [Borr85]. A copy resolution of 6 lp/mm has been achieved. Homogeneous lens arrays can be fabricated on large sheets of glass with dimensions as high as 8.5- x 11-inch. That would allow the copying of a normal paper format.

Fig. 5.6 – Coupling between two linear fiber arrays using refractive and diffractive lenses (After [Defo95])

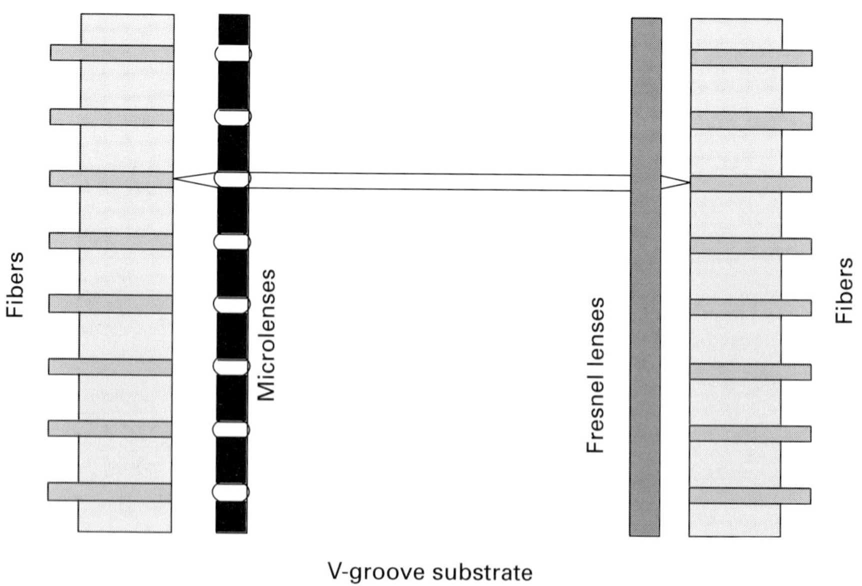

Fiber coupling

These lenses are also in use for micro-optical applications. In [Defo95] a linear array of lenses is used to collimate the outcoming light of 8 single mode fibers. The principle is shown in fig. 5.6. The task is to couple two linear single mode fiber arrays (8 fibers per array) with the help of microlenses. Light from the left fiber array is collimated by refractive microlenses. For incoupling into the second fiber array diffractive Fresnel lenses are used. In this experiment the overall coupling losses for every fiber were in a range of about 10 dB to 11 dB. The pitch error of the lenses was found to be the limiting property of the setup. In general the pitch error in the refractive array was larger than in the diffractive lenses. But diffractive lenses suffered from low diffraction efficiency and crosstalk.

Integrated tunable receiver

Another application of photolytic microlenses has been reported in [Tong95]. An integrated tunable receiver has been built using a hybrid approach of guided wave optics, free space optics and electronics (fig. 5.7).

Fig. 5.7 – Tunable receiver using photolytic microlenses (After [Tong95])

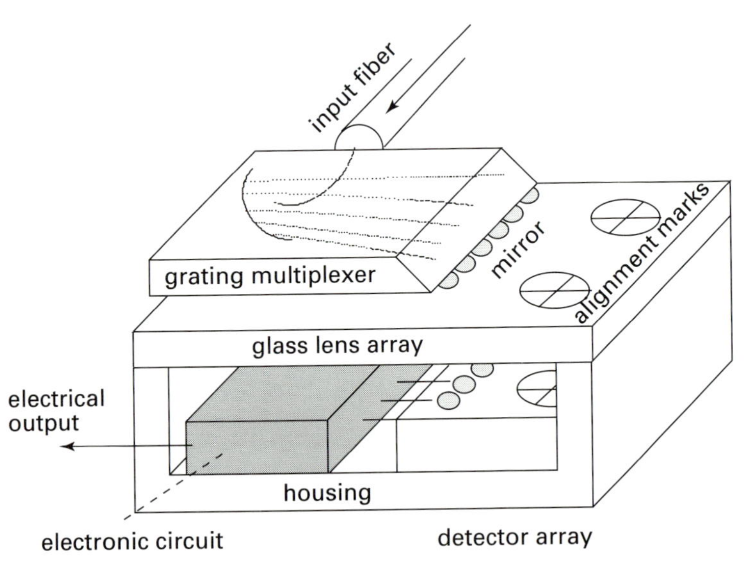

Signals in a wavelength range between 1537 to 1569 nm arrive in an input multimode fiber. A waveguide grating multiplexer images the light via a 45° tilted mirror and refractive microlenses onto a detector line. 32 channels are available with a channel separation of ≈ 1 nm and a tuning time of ≈ 10 ns. The microlenses have a diameter of 100 µm and a center to center spacing of 140 µm. The detectors have a size of 75x75 µm^2.

5.2.2. Application of anisotropic etching

Mechanical mounts

Given by the very high etch depth (as high as 2 mm) these substrates can be useful for holding structures. This is necessary eg. as mounting plates for microopitcal components or as fiber holders. From the given diameter tolerances it is obvious that eg. fiber holder arrays can only be made for multimode fibers with large core diameter. The global shrinkage of the substrate which influences the overall pitch accuracy has also be taken into account.

Fiber stub array

Instead of making holder structures for fibers it is more accurate to etch fibers directly into the substrate. This is shown schematically in fig. 5.8 [Goos92]. The fabrication is straightforward. Via a chromium mask with small opaque disks the 1.6 mm thick substrate is illuminated. The etch from both sides was timed such that a 170 µm thick substrate rests between the fiber stubs. There is an asymmetry between the stubs on both sides due to the irradiation from only one side. The upper side has fiber stubs with a diameter of 50 µm and 730 µm long. On the opposite side the diameter is 65 µm and the length of the stubs is 670 µm. Since slight etching of nonexposed regions occurs the pillars have a taper angle of 3°. But the side walls and the fiber ends are roughened. For that reason polishing of the stub ends is necessary. The resulting fiber stub is a multimode guide. It can be used for direct access to active elements rather than microlens arrays.

Fig. 5.8 – Fiber stub array directly etched into photosensitive glass
(After [Goos92])

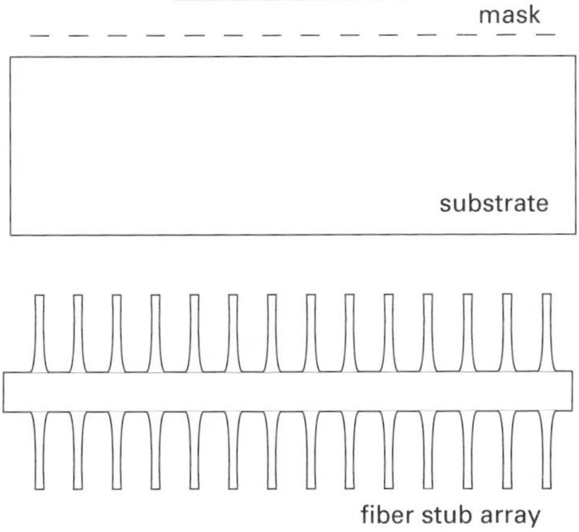

The etch process roughens the side walls, therefore the etching is not suitable to fabricate microstructures for light deflection.

References for Chapter 5

[Bore85]   N. F. Borrelli, D. L. Morse, H. Bellman, and W. L. Morgan: "Photolytic technique for producing microlenses in photo-sensitive glass", Applied Optics, Vol. 24, No. 16, p. 2520 - 2525, 1985.

[Bore88]   N. F. Borrelli and D. L. Morse: "Microlens arrays produced by a photolytic technique", Applied Optics, Vol. 27, No. 3, p. 476 - 479, 1988.

[Corn89]   Corning product information "FOTOFORM: a material and a capability", 1989.

[Defo95]   Y. Defosse, L. Bonnel, P. Gravey: "Diffractive micro-optical elements forlinear arrays of singlemode fibers", in proc. of 'procédés et modèles pour la micro-optique passive', Metz, 11/12 april, p. 137 - 140, 1995.

[Goos92]   K. W. Goossen and J. A. Walker: "Monolithic optical fiber stub array", Optics Letters Vol. 17, No. 5, p. 381 - 383, 1992.

[Scho88]   Schott product information "FOTURAN: Feinstrukturierte Bauteile aus Glas und Glaskeramik", Nr. 4844/2 d, 1988.

[Tong95]   F. Tong, K. Liu, C. S. Li, and A. E. Stevens: "A 32-channel hybridly-integrated tunable receiver", in proc of 21st Eur. Conf. on Opt. Comm. (ECOC'95- Brussels), p. 203 - 206, 1995.

Gradient Index Components

The term 'graded index' or 'gradient index' denotes to those materials where the refractive index varies continuously inside the medium. This has already been mentioned in chapter 4 for a gradient index waveguide. Now the index change penetrates deeper into the substrate. Useful applications have been found in the fabrication of microlenses and fibers for imaging.

6.1. Fabrication process

6.1.1. Procedure

The process used for index variation is diffusion. Commonly used substrates include glass, plastics and $LiNbO_3$. fig. 6.1 shows the fabrication procedure for microlenses. During diffusion also a small volume growth of the material occurs because the indiffusing ions have slightly larger diameter.

This technique is used especially for glass substrates and $LiNbO_3$. A slight index change takes place where ions such as Ag^+, K^+, Tl^+ diffuse into the substrate and replace lighter ions such as Na^+. Fig. 6.1 shows the principle ion diffusion for a glass substrate and the sodium-silver exchange process. In a first step the substrate is covered with a metal layer on both sides (typically alumnium or titanium). It is very important to have a good adhesion of the metal layer to the substrate which will resist the hot molten salt to several hours. Typical salt materials are $AgNO_3$ (melting point: 208°C), KNO_3 (339°C), and $TlNO_3$(230°C). During diffusion a slight index change takes place which is typically in the order of $\Delta n \approx 0.1$ for Tl^+ ions, $\Delta n \approx 0.08$ for Ag^+ ions ,and $\Delta n \approx 0.02$ for K^+ ions. Tl^+ and Ag^+ions are also used for the fabrication of multimode waveguides, K^+ ions for singlemode waveguides. Typical diffusion times for waveguides are in the order of a few hours. Microlenses often need more than 100 hours.

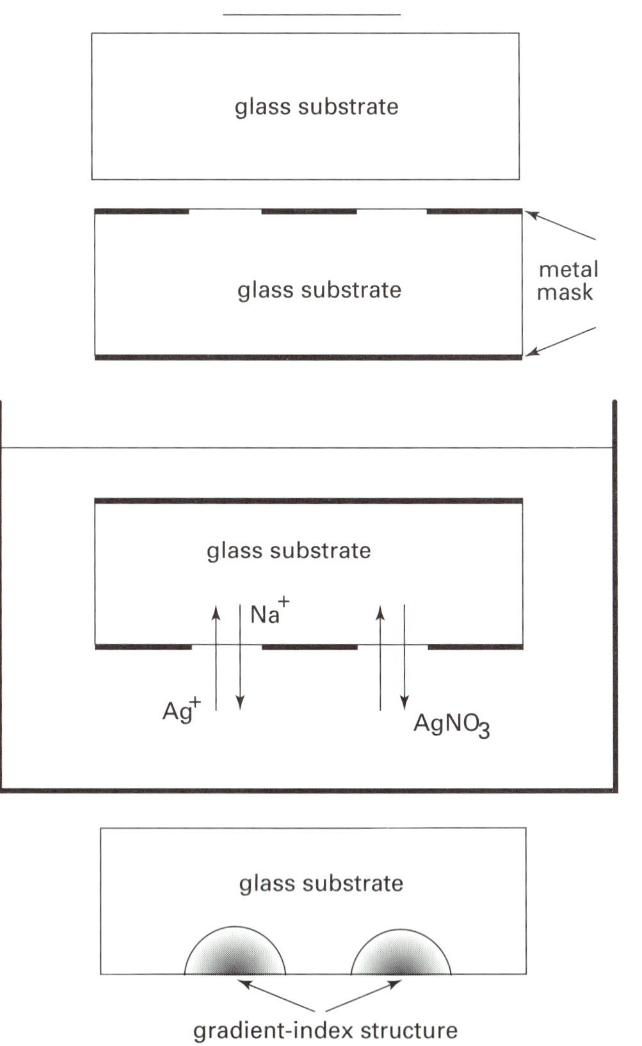

Fig. 6.1 – Thermal ion diffusion using the Na-Ag exchange process (After [Test94])

6.1.2. Modelling the process

The primary effect of the diffusion is changing the concentration of the diffusant in the substrate. Most interesting for the optics however is the variation of the refractive index which is caused by the concentration change. Therefore the modelling is divided into a first section about the diffusion itself and the concentration change and a second one about the related optical properties.

Diffusion parameters
The information about the final index distribution can be obtained from modelling the diffusion process: The diffusion phenomena in one dimension is expressed by

$$\frac{\delta C}{\delta t} = \frac{\delta}{\delta x} (D \frac{\delta C}{\delta x})$$

(6.1)

where C is the concentration of the diffusing substrate, x the space coordinate measured normal to the section, and D is the diffusion coefficient. The diffusion equation depends on the behaviour of D. Different diffusion environments show strongly different D. The easiest model assumes D to be constant. Then Equation 6.1 can be written

$$\frac{1}{D} \frac{\delta C}{\delta t} = \frac{\delta^2 C}{\delta x^2}$$

(6.2)

This equation is a common parabolic diffusion equation for which many numerical solutions exist. These solutions can be superimposed to give new useful solutions for different applications. In [Koik92] this model is used to describe diffusion in polymers. In general every diffusion process has its specific behaviour for D. Two other models will be mentioned that refer to different processes. An ion exchange between Tl^+ and Na^+ in glass makes use of the so-called k-model which describes exponential concentration dependancy of D

$$D = D_0 \, e^{k \frac{c}{c_0}}$$

(6.3)

C_0 is the saturation concentration of the material. k describes the degree of nonlinearity of the process. In the third model D is given by

$$D = \frac{D_i}{1-\alpha C} \text{ with } a = 1 - \frac{D_i}{D_0}$$

(6.4)

D_i is the self-diffusion coefficient for the indiffusion and D_0 for the out-diffusion. $\alpha = 0$ gives equation 6.2. This model can be used for the ion exchange between Ag^+ and Na^+ in glass. In the literature α is given in a range of 0.5 to 0.9 [Test94].

Optical Parameters
For optical applications the relationship between concentration distribution and refractive index distribution n is important. It is given by the Clausius-Mossotti formula

$$(n + n_2)^2 = n_2^2 \frac{1 + \frac{2}{3}\rho C}{1 + \frac{1}{3}\rho C} \tag{6.5}$$

where n is the change in the refractive index, n_2 is the index of the substrate, and ρ is a material specific parameter which often has to be determined experimentally.

For microlenses, where the index variation in axial and radial direction changes, the index distribution can be represented by

$$n(r,z) \approx n(0,0) \sqrt{1 - g^2 r^2 - \frac{2\, g\, n^2(0,0)\, \Delta}{d} z^2} \tag{6.6}$$

r radial distance
z axial distance
$n(0,0)$ maximum index at surface
d diffusion length
Δ $= 1 - n_2/n(0,0)$
g $= \sqrt{(2\Delta)}/r_{max}$
n_2 refractive index of substrate

Typical values for the Tl^+/Na^+ process are

$n(0,0)=0.1$, $\Delta=0.05$, $d = 400$ μm, $r_{max}=500$ μm, $n_2=1.5$, $g=0.63$ mm^{-1} [Iga84].

In [Bähr94] a relationship is given that combines the size of diffusion mask, diffusion time, focal length and numerical aperture. It is assumed that a plane wave passes through a thin lens and a phase shift $\Delta\phi$ is produced.

$$\Delta \phi = \phi_{max} - \frac{\pi x^2}{\lambda f},$$
$$\phi_{max} = \frac{\pi r_L^2}{\lambda f} \tag{6.7}$$

This is the phase shift produced by a parabolic lens of focal length f for a wavelength λ. The maximum phase shift is normalized to zero at the lens edge $x = r_L$.

ϕ_{max} can be expressed as

$$\phi_{max} = \frac{2\pi}{\lambda}\, n\,(0,0)\, \delta z \beta \tag{6.8}$$

Gradient Index Components

where δz is the ion-penetration depth and β is a parameter that characterizes the ion distribution along the z axis. For a linear decay, e. g., β can be shown to be 0.5. If the lens radius is expressed as a sum of the mask radius r_m and the ion penetration depth δz the focal length is

$$f = \frac{1}{2\, n\,(0,0)\,\beta} \left(\frac{r_m^2}{\delta z} + 2 r_m + \delta z \right) \tag{6.9}$$

The numerical aperture can be given as

$$NA \approx \frac{r_L}{f} = \frac{2\, n\,(0,0)\,\beta\,\delta z}{(r_m + \delta z)} \tag{6.10}$$

If a small mask aperture can be assumed, i. e. $r_m \ll \delta z$, equation 6.10 reduces to

$$NA \approx 2\, n\,(0,0)\,\beta \tag{6.11}$$

This means that the NA is determined by the maximum index change at the surface and the ion distribution parameter. Table 6.1 lists a set of values resulting from equation 6.9 and 6.10 for the Ag^+/Na^+ ion exchange process [Bähr94].

Table 6.1

maskØ/µm	lensØ/µm	f/µm	NA
40	42	590	0.036
60	99	560	0.089
100	116	950	0.061
120	125	1230	0.051
140	138	1380	0.050

6.1.3. Variations of the process

Field assisted ion exchange
The process can be changed if an electrical field is applied along the diffusion direction. The diffusion time for ion diffusion into glass is in the order of 100 hours. During this long time period a diffusion environment with constant temperature has to be provided. For plastic materials diffusion is much shorter (< 1 hour). But also for ion diffusion the time can

be significantly reduced to about 1 hour if an electrical field is applied along the diffusion direction of the ions. In this case the mobility and penetration depth of the ions is much higher. The diffusion time is reduced to about 30 minutes. Fig. 6.2 shows the setup for field assisted ion exchange. In addition the electrical field causes the ions to penetrate deeper into the substrate and the NA becomes higher. Fig. 6.3 shows a cut through the plane of the lens axes. The spherical index distribution can be seen. Fig. 6.4 illustrates the difference in phase shift between ordinary and field assisted diffusion.

In [Bähr95] a microlens with NA = 0.2 is reported which has been fabricated by the field assisted diffusion. The other parameters are:

- mask aperture: 2 μm
- external field: 125 V/mm
- temperature: 320°C
- diffusion time: 60 min
- lens diameter: 350 μm
- focal length: 843 μm
- n(0,0): 0.08

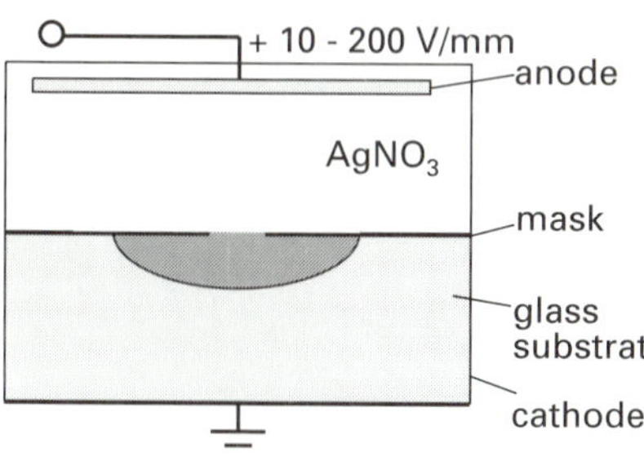

Fig. 6.2 – Schematic diagram of field assisted ion exchange at about 300°C (After [Bähr94]).

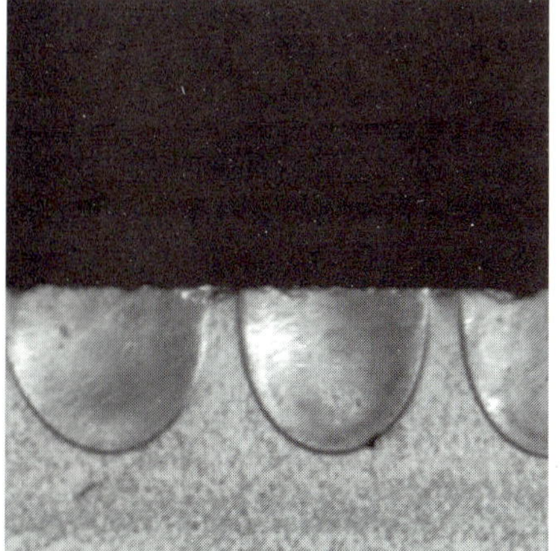

Fig. 6.3 – A cut through the plane of the lens axis (courtesy Jochen Bähr [Bähr95])

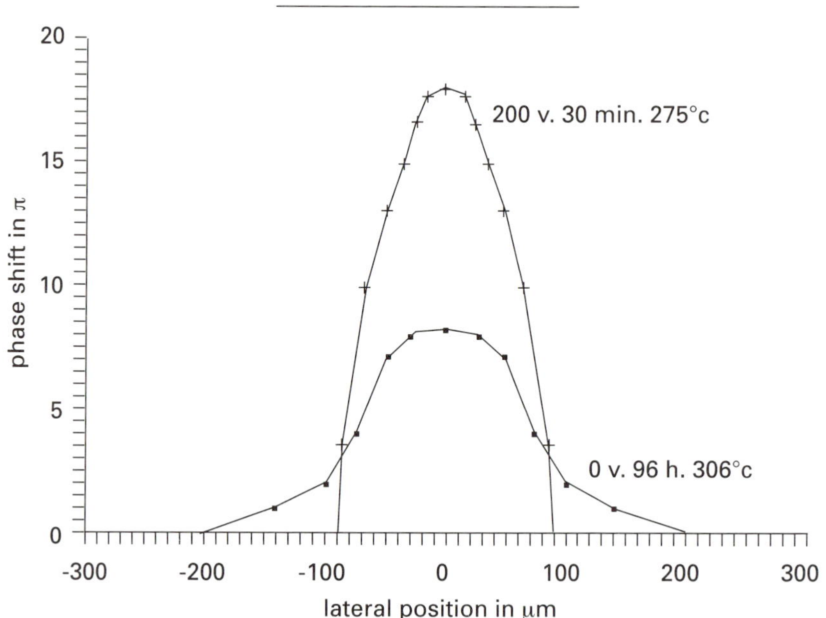

Fig. 6.4 – Phase profiles for a lens with and without field assisted diffusion (After [Bähr95])

Astigmatic gradient index lenses

In [Sinz95] the fabrication of astigmatic gradient-index microlenses is described. The goal is for collimating an edge-emitting laser diode beam which has two different diverging angles for the lateral dimensions x and y. This is caused by the rectangular shape of the output window of a laser diode. Collimating such a beam profile needs astigmatic collimation optics with two different focal length for the two perpendicular lateral extensions (fig. 6.5). In table 6.2 several parameters of astigmatic gradient-index lenses are listed. The mask aperture is elliptical with different diameters in x and y direction. The resulting lens has also two different focal lengths and two numerical apertures for x and y. The ratio f_1/f_2 must be chosen such that it is adapted to the parameters of a given laser diode. An alternative solution for this problem uses surface microlenses fabricated by mass transport [Liau95]. The lens side facing the laser diode has a cylindrical shape. The opposite side is used for coupling into fibers and is almost spherical. Mass transport is a melting process (see chapter 7) but with very high temperatures in semiconductor materials like GaP.

Table 6.2

maskØ/µm		f/µm		NA		ratio f_1/f_2
		f_1	f_2			
80	40	730	390	0.064	0.089	1.87
100	10	1140	330	0.047	0.079	3.5
200	50	2980	566	0.03	0.088	5.26
300	30	8230	430	0.015	0.085	19.1

Fig. 6.5 – Interferogram of astigmatic gradient index lenses (courtesy Stefan Sinzinger [Sinz95])

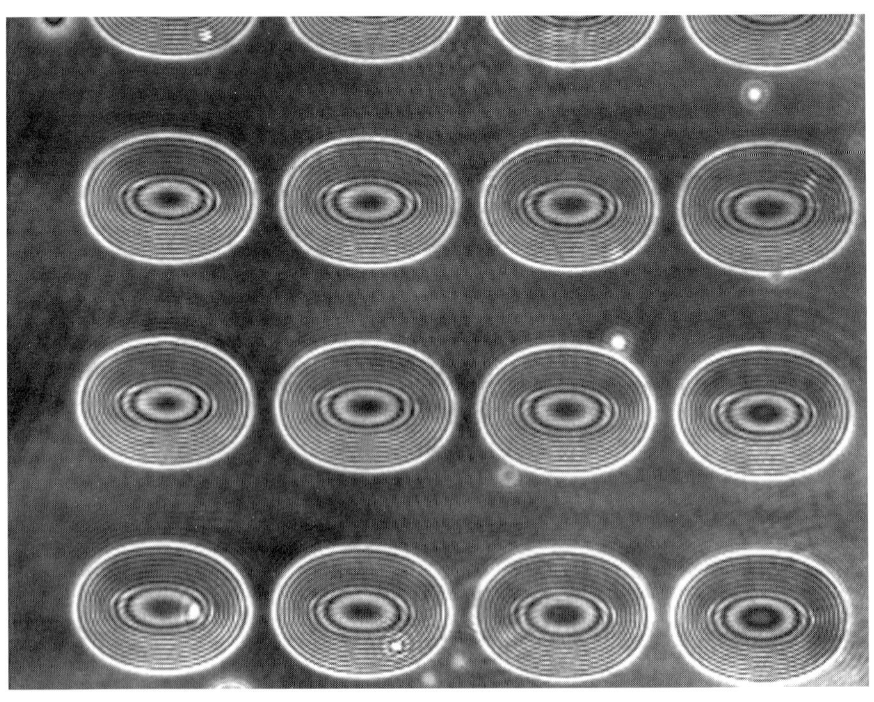

Radial GRIN lenses

In the case of a radial gradient the GRIN material can be fabricated into a rod shape. The index profile of such a rod is shown in fig. 6.6. The refractive index is given as

Gradient Index Components

$$n(r) = n_2 \left(1 - \frac{A}{2} r^2\right) \tag{6.12}$$

where n_2 is the substrate index and A is a gradient constant. The numerical aperture of such a lens can be calculated from [Self94]:

$$NA^2 = \frac{A \left(n_2 \left(1 - \frac{A}{2} r^2\right)\right)^2 r_0^2 (1-R^2)}{1 - R^2 \sin \phi} \tag{6.13}$$

with the parameters:

r_0 lens radius (in mm)
r distance from center axis (in mm)
$R = r/r_0$ normalized distance from center axis
$\sqrt{A}$ index gradient constant (in mm^{-1})
n_2 substrate index
ϕ angular distance from lateral axis

The numerical aperture is maximal at the center of the lens, and decreases with the distance from the optical axis. Therefore an effective diameter of 60 % to 70 % is suggested for best optical performance

Table 6.3 lists several parameters of different types of gradient-index rod lenses for a wavelength of $\lambda = 633$ nm taken from [Self94]. These lenses are fabricated using the Tl^+/Na^+ ion exchange process.

Table 6.3

NA	diameter/mm	n_2	$\sqrt{A}$/mm^{-1}
0.37	1.0	1.5637	0.499
0.37	2.0	1.5637	0.247
0.46	1.0	1.6075	0.608
0.46	1.8	1.6075	0.339
0.46	2.0	1.6075	0.304
0.46	3.0	1.6075	0.206
0.6	1.8	1.6576	0.430

Fig. 6.6 – Typical refractive index profile of a GRIN rod

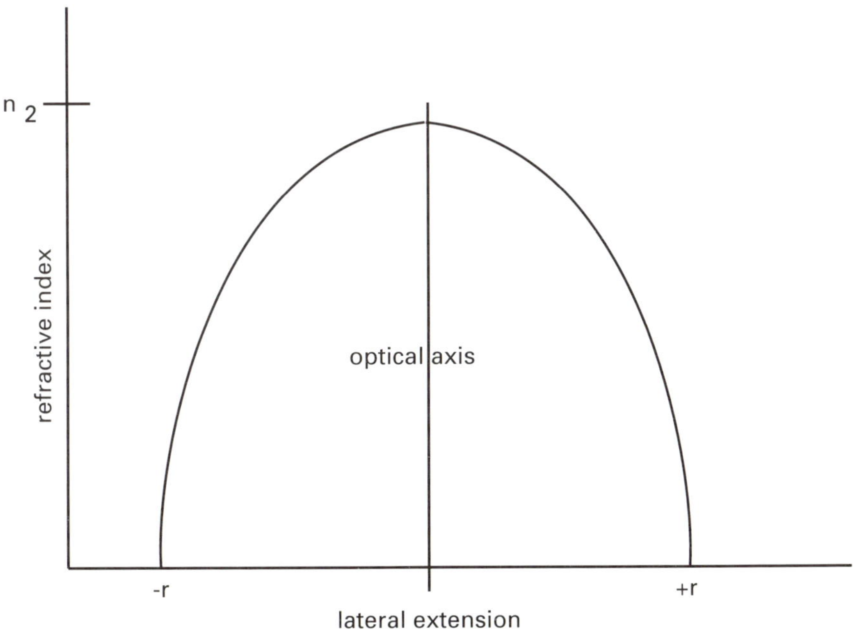

Fig. 6.7. illustrates the difference between a graded index rod lens and a conventional lens: The imaging effect is achieved in case of the conventional lens by refraction at the lens surfaces, in case of the rod lens the light is bent in the direction of increasing refractive index, which means towards the center of the rod.

The same principle works also for guiding light in a gradient-index fiber, which can be regarded as a very long GRIN rod with a small diameter. (see fig. 6.8.) In a step index material light is guided by total internal reflection. This leads to a significant path difference of rays (see eg. fig. 4.2). In a GRIN rod the rays passing through the center of this rod lens propagate more slowly than those further away from the optical axis. Due to the different pathlength of the rays they arrive approximately at the same point.

Fig. 6.7 – Illustration of imaging for a conventional lens and a gradient-index rod lens

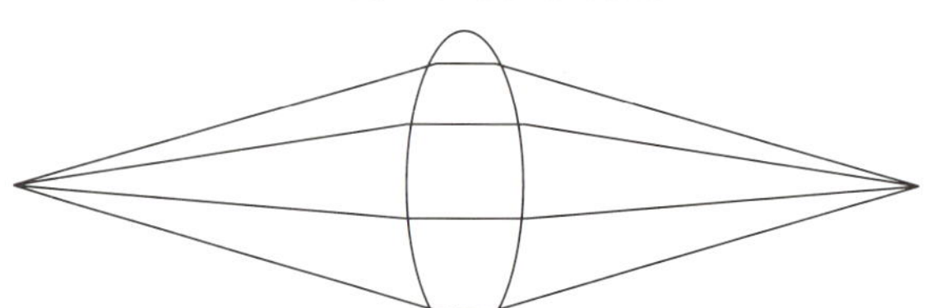

Conventional lens

Gradient index rod lens

Fig. 6.8 – Gradient index fiber versus step index fiber

step index fiber

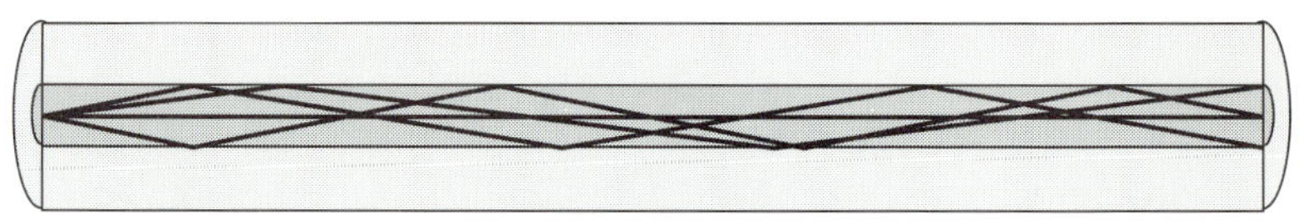

Gradient index fiber

6.2. Applications: optical interconnects using gradient index lenses

Gradient index optics nowadays is commercially available as arrays of microlenses, as single rod lenses and as gradient index fibers. Various important system aspects for free space optics using GRIN elements have been shown in the literature.

Fig. 6.9 – Concept of stacked planar optics (After [Iga84])

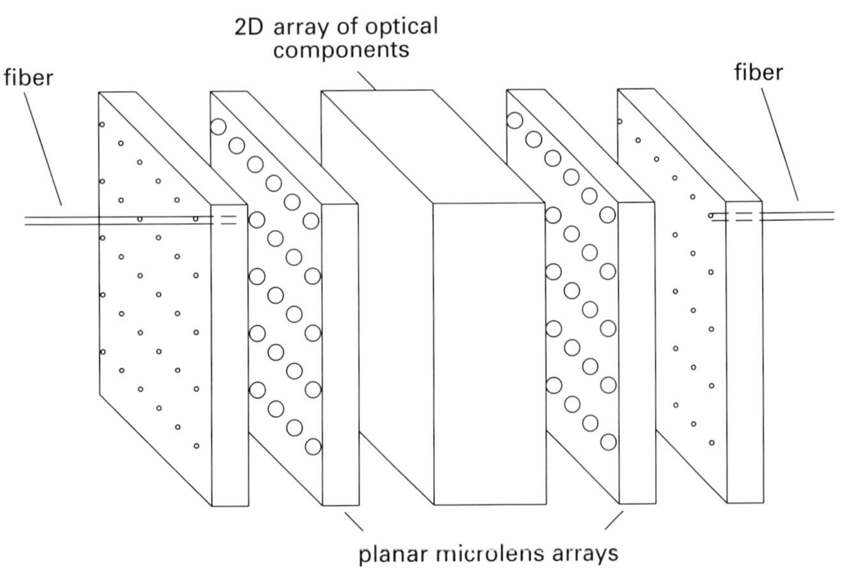

Stacked systems

Two-dimensional microlens arrays have led to the idea of stacked planar optics [Iga84]. The concept is illustrated in fig. 6.9. All optical components such as fiber arrays and microlens arrays are aligned to each other plane by plane. This architecture is designed for three-dimensional free space optical concepts.

In [Mizu95] a compact version of a stacked system has been published using a so-called guide frame assembly method (fig. 6.10). In a stable housing frames for positioning optical elements such as single lenses, microlens arrays or fiber arrays are included. Spacer frames allow exact spacing between neighbouring frames. The tolerances for a compact fiber array/microlens assembly are given below.

Two-dimensional optical fiber array
Size   7.0 mm x 7.0 mm x 2.0 mm
Array size   8 x 8
Array pitch   250 μm
ptich error (average)   3 μm

Gradient Index Components

Planar microlens array
Size   4.5 mm x 4.5 mm x 1.8 mm
Array pitch   250 μm
Focal length   600 μm

Guide frame and spacer frame
Size (guide frame)   20 mm x 20 mm x 1 mm
Size (spacer frame)   20 mm x 20 mm x 2.34 mm
Right-angle tolerance   0.05°
Parallelism   3 μm
Geometrical tolerance   5 μm

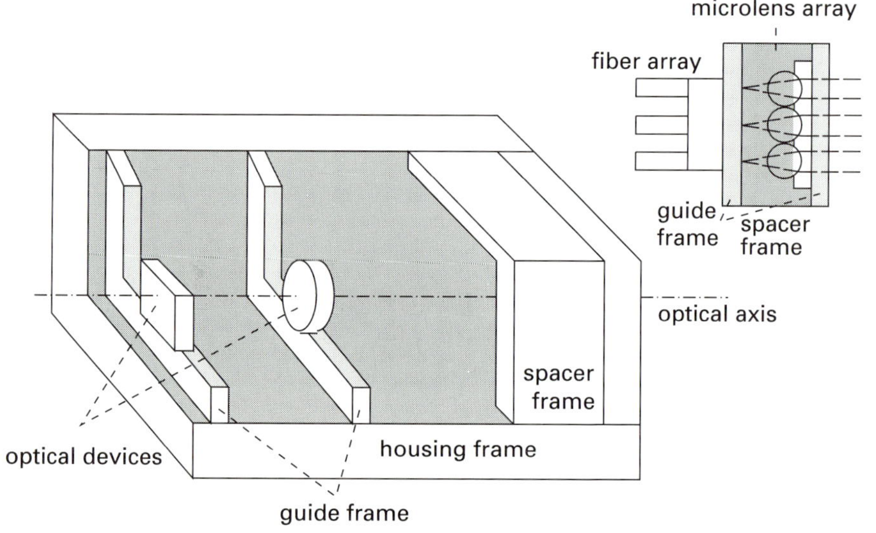

Fig. 6.10 – Guided frames using the concept of stacked planar optics
(After [Mizu95])

Fiber- lens connector

For the fabrication of a stable and easy alignable fiber-lens connector a
monolithic fabrication method has been reported in [Sasa92].

Fig. 6.11 – Plug-in self-aligning scheme for microlenses and fibers
(After [Sasa92])

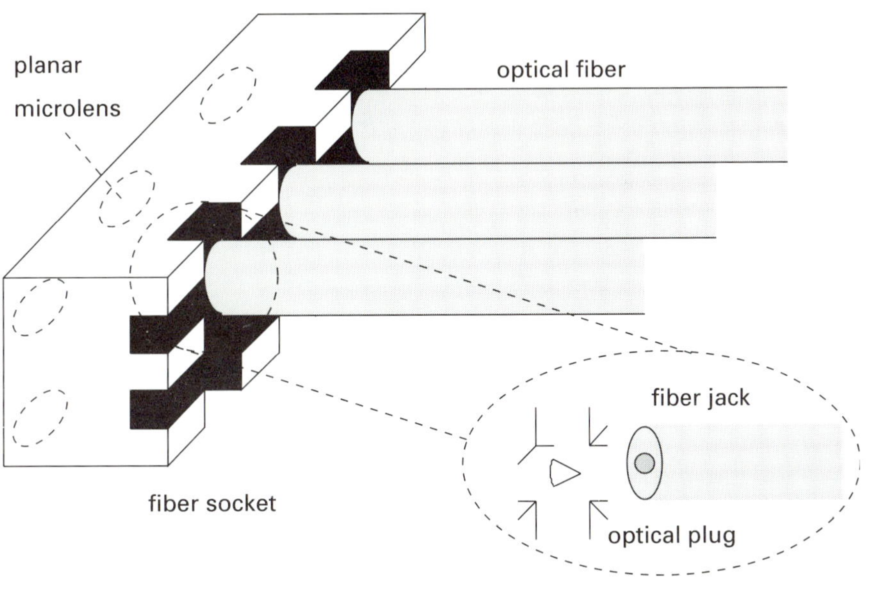

On the rear side of an ion-diffused glass microlens array a saw is used to cut fiber socket grooves as shown in fig. 6.11. The grooves are filled with negative photoresist and the lens side is then illuminated by a parallel beam of an argon laser. During irradiation small cones (optical plugs) are formed in the grooves by the light intensity of the focal spot of the lenses. After development these cones fit into fiber jacks that have been created by etching the core of the fibers. Typical height of the plugs is in the range between 20 μm and 50 μm with a diameter of about 20 μm. With this technique the fiber core can be aligned relatively to the focal plane of the lenses. To fix the fiber jacks and also to reduce the light scattering at the rough surface of optical plugs, fiber jacks and fiber sockets are all covered by UV stiffed resin.

Optical bus interconnection

Using the concept of stacked planar optics, a board to board optical interconnection scheme has been proposed in which glass rods with perpendicular gaps formed by slicing are aligned and fixed on a mother board [Hama91].

Gradient Index Components

Fig. 6.12 – Optical bus interconnection system with gradient index rod lenses (After [Hama91])

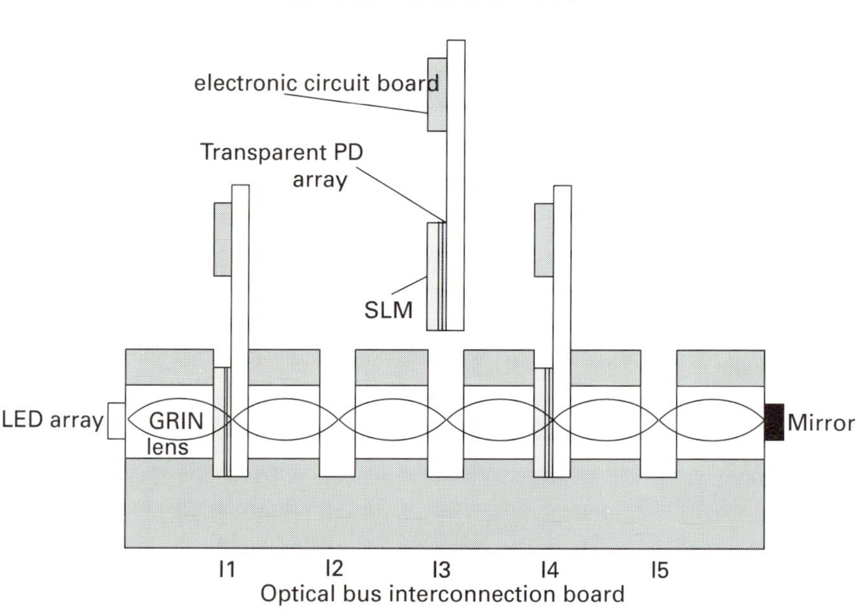

The gap pitch is determined such that conjugate image planes are formed at the same positions in all gaps. The inserted boards have transparent photodetectors and spatial light modulators and are index matched in the gaps. A two-dimensional LED array at the one end serves as input and a mirror reflects the light at the other end (fig. 6.12). Inserting conventional beamsplitters and attachment of fiber bundles has also been reported [Hama94].

Static dataplane interlace

For miniaturization of optical computing systems using symbolic substi-tution algorithms microprisms with small vertex angles are useful. In [Mois95] mechanical fabrication of microprisms is proposed. First a glass master with the prisms is made by polishing techniques having a prism angle of 8° and the exact lateral dimensions. The glass master is then heated and embossed in a PMMA substrate and cooled down. If necessary this procedure can be repeated several times to produce an array of microprisms. This PMMA substrate can serve as a negative master structure for the final prism array. For this purpose, it is filled with UV-curing glue and covered with an array of planar microlenses.

Four lenses each are aligned under a pair of microprisms as shown in fig. 6.13 (see also photograph, fig. 6.14). Then the glue is cured by UV light and the negative is separated from the system. The focal length of the lenses is adapted to the thickness of the glass substrate (f = 1500 μm, in glass). The lens diameter is 200 μm and the pitch is 300 μm.

The experimental data from this setup are as follows

size of output plane	400 μm x 400 μm
number of data channels	8 x 8
size of rectangles/circles	40 μm x 40 μm/ 40 μm diameter
separation of data channels	10 μm
S/N-ratio	17:1 (λ = 450 nm), 11:1 (λ = 633 nm)
light efficiency (excl. Fresnel losses)	89%
peak/channel homogeneity	7%/4%

In analogy the system can also be used in the other direction to perfom a multiplex operation.

Fig. 6.13 – Interlace of data planes using a system of microlenses and microprisms (After [Mois95])

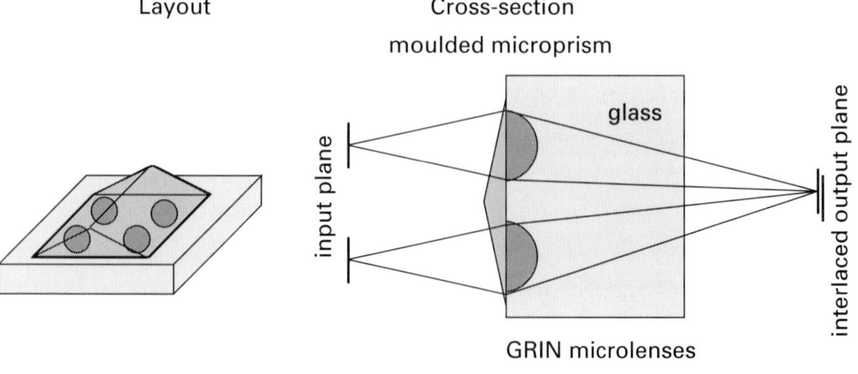

Fig. 6.14 – Photograph of the microlens array covered with microprisms
(Courtesy Jörg Moisel, [Mois95])

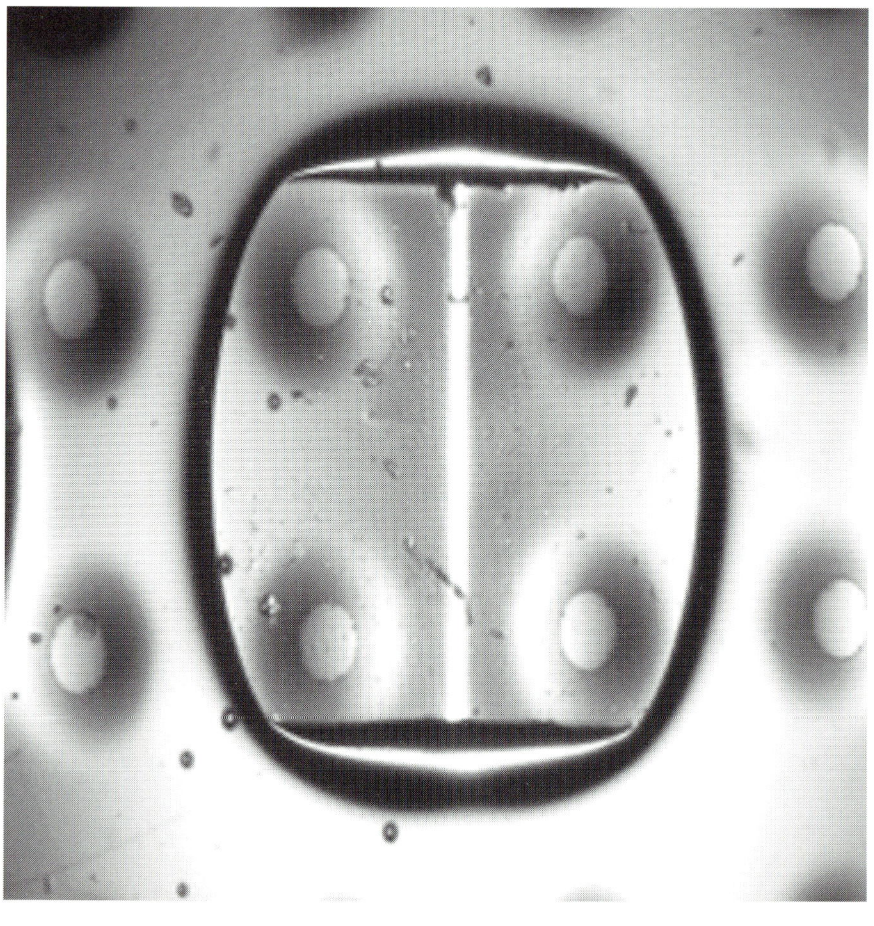

Refractive planar optical system

This principle has been extended for use in a planar approach (see chapter 3.4) but using refractive optics instead of diffractive (fig. 6.15) [Ecke95]. The system uses only a single lens layer where lenses are situated on both sides of the substrate. In addition one side is used for gluing the microprisms. The other side can be used for active elements and mirrors. Light path folding is applied to make the system very compact. The microlenses are used off axis which will induce aberrations.

Fig. 6.15 – Planar optics with refractive microlenses and microprisms
(After [Ecke95])

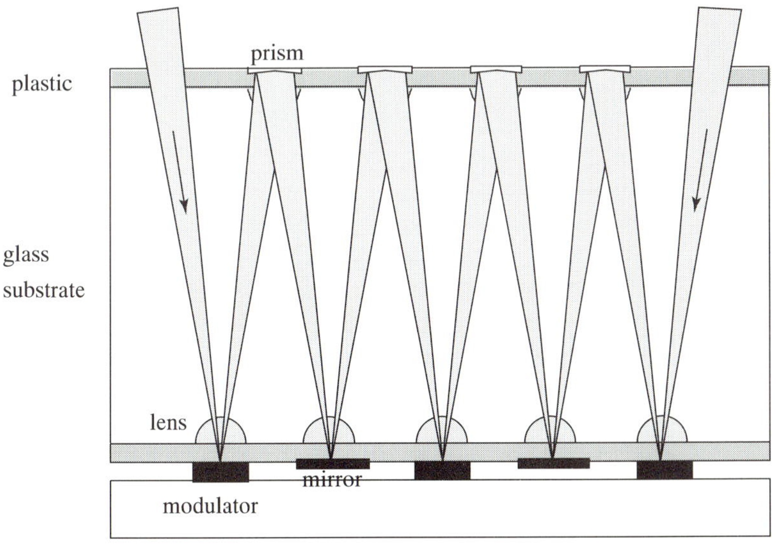

Fig. 6.16 – Structure of a liquid crystal microprism array (After [Hira95])

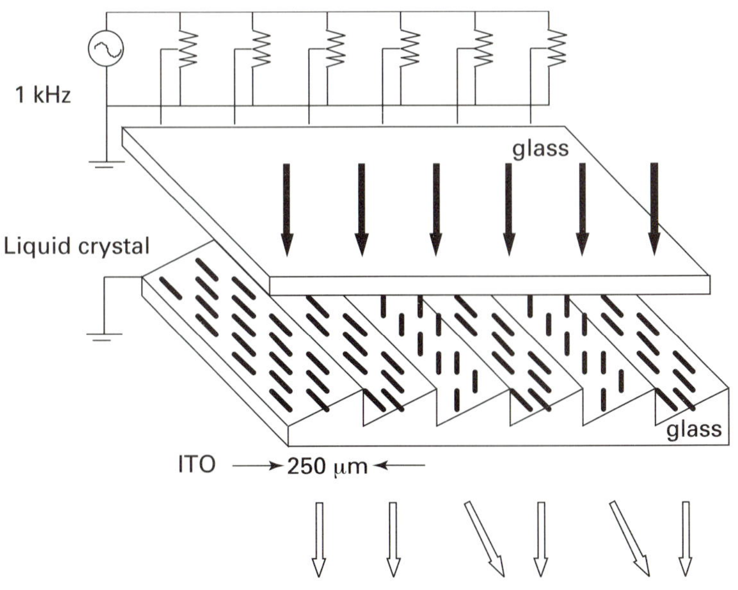

Dynamic switching using liquid-crystal microprism arrays

Dynamic interconnects with microprisms can be fabricated if microprisms are brought in contact with liquid crystals (fig. 6.16.) [Hira95]. An indium tin oxide transparent electrode is deposited on the microprisms fabricated by diamond milling. Liquid crystals are filled between the microprism array and a glass cover plate. If a voltage is applied the beam is not deflected. Without a voltage the maximum deflection angle appears. In the range in between the angular deflection can be varied according to the applied external voltage. The prisms can have an apex angle between 10° and 40°. The deflection angles corresponding to the applied voltage can be as high as 10°. The voltages are smaller than 2.8 V. By using two crossed liquid crystall microprism arrays beam deflection in two orthogonal directions is possible (fig. 6.17.). The overall pitch in the fiber array, microlens array and microprism array is 250 µm.

Fig. 6.17 – Interconnection by two crossed liquid crystal microprism arrays (After [Hira95])

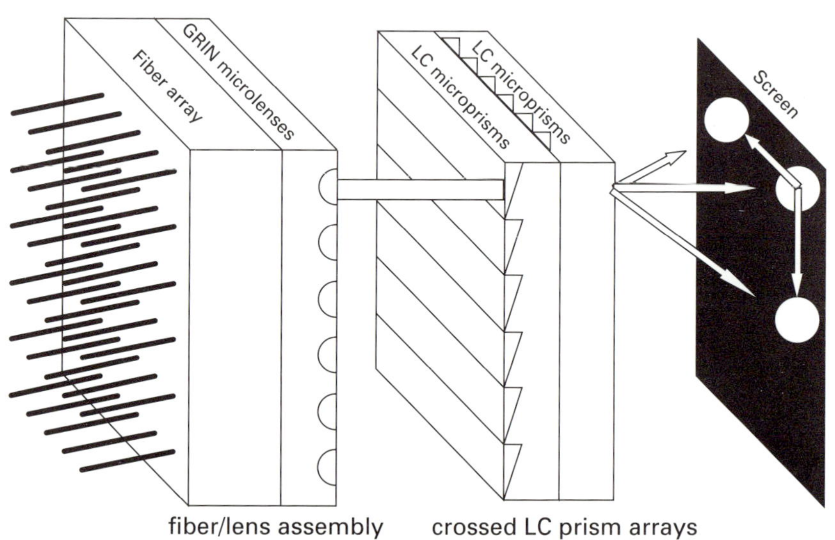

Three-dimensional free space data transcription

In [Kirk96] free space optical data transcription between arrays of opto electronic emitter-receiver elements has been reported. The elements used in this experiment were PnpN optical thyristors [Knüp95] and the optical system is shown in fig. 6.18 and fig. 6.19. Two 0.2 pitch gradient refractive index rod lenses are used to image the left side device plane onto the right side device plane. The GRIN rods have a diameter of 5 mm and a length of 30 mm. A beam-splitter provides an optical port to each plane. The optical system has a field of view of 2.5 mm. Due to the fact that photo thyristors are Lambertian sources the plane to plane transmittance is less than 1 %. This limited the data transmission rate to 1 MHz.

Fig. 6.18 – Optical free space interconnection scheme using GRIN rod lenses (After [Kirk96a, Kirk96b])

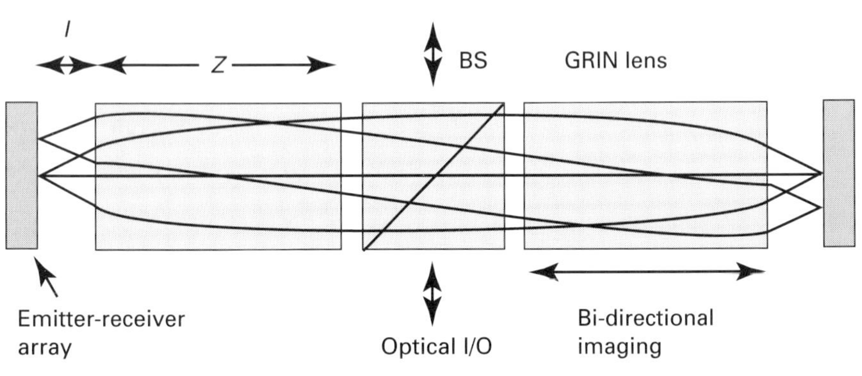

Fig. 6.19 – Photograph of the above system (Courtesy Andy Kirk and Hugo Thienpont, [Kirk96b])

110

References for Chapter 6:

[Bähr94]   J. Bähr, K.-H. Brenner, S. Sinzinger, T. Spick, and M. Testorf: "Index-distribution planar microlenses for three-dimensional micro-optics fabricated by silver-sodium ion exchange in BGG35 substrates", Applied Optics, Vol. 33, No. 25, p. 5919 - 5924, 1994.

[Bähr95]   J. Bähr, K.-H. Brenner: "Relisation of 0.2-N.A. microlenses by field assisted Ag-Na ion exchange", in Microlens Arrays, Vol. 5, 1995 EOS Topical Meetings Digest Series, p. 121 - 124, 1995.

[Ecke95]   W. Eckert, K.-H. Brenner, C. Passon: "Planar integration of free-space micro-optical systems with refractive elements", Optical Computing, Series Number 139, p. 263 - 266, IOP Publishing, London 1995.

[Hama91]   K. Hamanaka: "Optical bus interconnection system using Selfoc lenses", Optics Letters, Vol. 16, No. 16, p. 1222 - 1224, 1991.

[Hama94]   K. Hamanaka, K. Nakama, D. Arai, Y. Kusuda, T. Kishimoto, and Y. Mitsuhashi: "Integration of free-space interconnects using selfoc lenses: optical properties of a basic unit", in Technical Digest of Int. conference on optical computing, p. 227 - 228, Edinburgh, 1994.

[Iga84]   K. Iga, Y. Kokubun, and M. Oikawa: "Fundamentals of micro-optics", Academic Press, Tokyo 1984.

[Kirk96b]   A. Kirk, A. Goulet, H. Thienpont, N. McArdle, K.-H. Brenner, M. Kuijk, P. Heremans, and I. Veretennicoff: "A compact optical imaging system for arrays of optical thyristors", to appear in Applied Optics, 1996.

[Kirk96]   A. G. Kirk, H. Thienpont, A. Goulet, P. Heremans, G. Borghs, R. Vounckx, M. Kuijk, and I. Veretennicoff:"Demonstration of optoelectronic logic operations with differential pairs of optical thyristors", Photonics Technology Letters, Vol. 8, No. 3, p. 467 - 469, 1996.

[Knüp95]   B. Knüpfer, M. Kuijk, P. Heremans, R. Vounckx, and G. Borghs: "Cascadable differential PnpN optoelectronic switch operating at 50 Mbit/s with ultrahigh optical input sensitivity", Electron. Letters, Vol. 31, p. 485 - 486, 1995.

[Koik92]   Y. Koike in 'Polymers for lightwave and integrated optics', chapter 3, "Graded index materials and components", p. 71 - 104, ed. L. A. Hornak, Marcel Decker Inc., New York 1992.

[Liau95]   Z. L. Liau, J. N. Walpole, J. C. Livas, E. S. Kintzer, D. E. Mull, L. J. Missaggia, and W. F. DiNatale: "Fabrication of two-sided anamorphic microlenses and direct coupling of tapered high-power diode laser to single-mode fiber", IEEE Photonics Technology Letters, Vol. 7, No. 11, p. 1315 - 1317, 1995.

[Mils95]   T. M. Milster in 'Handbook of Optics', Vol. II, chapter 7, "Miniature and micro-optics", p. 7.13 - 7.16, ed. M. Bass, E. W. Van Stryland, D. R. Williams, W. L. Wolfe, McGraw Hill, New York 1995.

[Mizu95]   M. Mizukami, K. Koyabu, M. Fukui, and K. Kitayama: "Free-space optical module configuration using a guide-frame assembly method", Applied Optics, Vol. 34, No. 11, p. 1783 - 1787, 1995.

[Mois95]   J. Moisel and K.-H. Brenner: "Demonstration of a 3D integrated refractive microsystem", Optical Computing, Series Number 139, p. 259 - 262, IOP Publishing, London 1995.

[Oika90]   M. Oikawa, H. Nemoto, K. Hamanaka, and E. Okuda: "High numerical aperture planar microlens with swelled structure", Applied Optics, Vol. 29, No. 28, p. 4077 - 4080, 1990.

[Sasa92]   A. Sasaki, T. Baba, and K. Iga: "Put-in Microconnectors for alignment-free coupling of optical fiber arrays", IEEE Photon. Technol. Lett. Vol. 4, p. 908 - 911, 1992.

[Self94]   SELFOC Product Guide, NSG Europe, 1994.

[Sinz95]   S. Sinzinger, K.-H. Brenner, J. Moisel, T. Spick, and M. Testorf: "Astigmatic gradient-index elements for laser-diode collimation and beam shaping", Applied Optics, Vol. 34, No. 29, p. 6626 - 6632, 1995.

[Test94]   M. Testorf: "Analyse des Ionenaustauschprozesses in Glas im Hinblick auf eine Synthese refraktiver optischer Mikroelemente", Thesis Uni Erlangen-Nürnberg, 1994.

Melting of Photoresist

7.1. Fabrication process

Melting of any material results in a flow of the substance with respect to surface tension. If polymer materials are chosen then lithographic techniques can be applied to preform the resist. The melting technique with photoresist for microlens fabrication is widely used because good optical quality lenses can be obtained with less expensive equipment in a relatively fast production time.

Fig. 7.1 – Process of melting photoresist

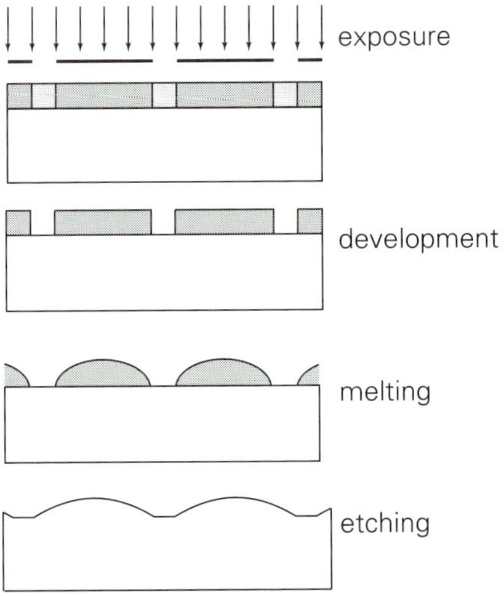

In the first step of fabrication a relatively thick layer of photoresist is spun on a substrate (eg. glass). Thickness in the range of 5 μm to 60 μm is used because the refractive lenses must have a certain sag. Data for a resist material suitable for thick layers are given in chapter 2.1.

The resist of known thickness is then exposed under a mask with opaque circles. The standard irradiation process can be modified in order to preform the resist better. For this purpose direct laser writing or greyscale masks can be applied as already described in chapter 3.1.3 (see eg. [Jay94]). Or as an alternative process for very long focal length mcirolenses, preshaping is done by moving a special photomask over the substrate during irradiation [Artz92, Gex96]. The subsequent development process yields either cylindrical islands or preshaped islands. These islands can be heated in an oven to about 140°C, when the resist melts and surface tension draws it into the shape of the cap of a sphere. These lenses may be used directly or may be processed further eg. etched into the substrate or for casting replicas in different materials.

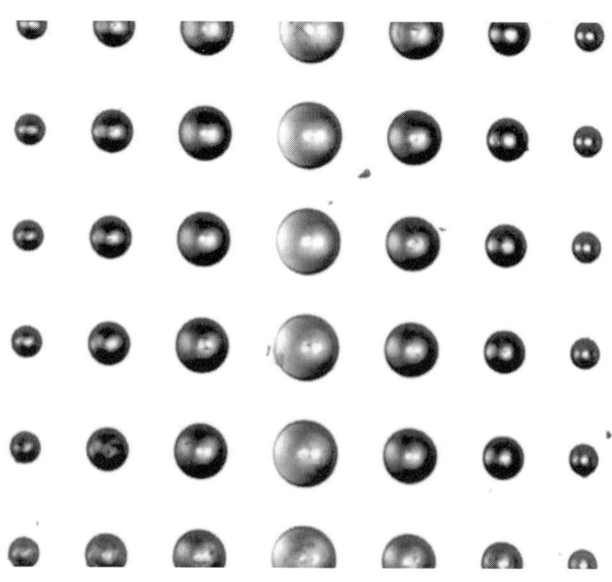

Fig. 7.2 – Photograph of a typical photoresist lens array (diameters are between 50 mm and 200 mm)

The thickness of the resist necessary to achieve a certain focal length can be calculated by assuming that the volume of a developed island and a melted lens is the same. Fig. 7.3 shows a typical surface lens.

Melting of Photoresist

Fig. 7.3 – Parameters describing a single surface lens

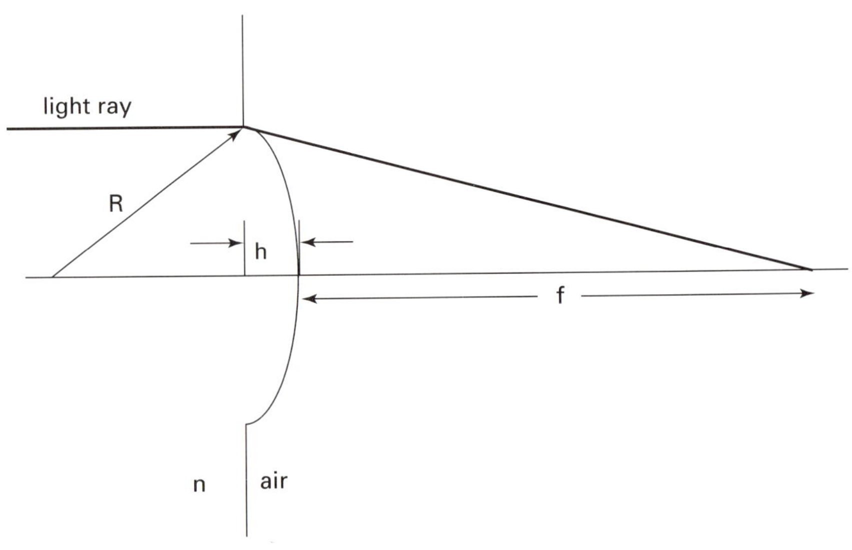

The paraxial focal length of the above lens is described by

$$f = \frac{R}{n-1} \tag{7.1}$$

and the heigth h is given by

$$h = R - \sqrt{R^2 - r^2} \tag{7.2}$$

The volume of a cylindrical island before and after melting is

$$V_{cyl} = T\pi r^2 \quad \text{and} \quad V_{lens} = \frac{1}{3}\pi h^2 (3R - h) \tag{7.3}$$

respectively. It follows for the necessary film thickness of the spin process

$$T = \frac{h}{6}\left(3 + \frac{h^2}{r^2}\right) \tag{7.4}$$

Table 7.2 lists various parameters for lenses with a fixed diameter of 100 µm and a refractive index of n = 1.6 [Daly90].

Table 7.2

f/µm	R/µm	h/µm	T/µm
100	60	27	14.8
200	120	10.9	5.3
400	240	5.5	2.6

Possible ranges of diameters for single resist layers are between 20 µm and 500 µm with F-numbers ranging between 1 to 2.5 [Daly90]. Higher F-numbers can only be achieved when the process is changed. The pre-shape procedure is a method for generating higher F-numbers [Jay94]. If a larger F-number is desired the lenses can be matched with index liquid with $n_{liquid} < n_{resist}$ and a glass cover. For small F-numbers a multiple resist layer technique can be used. Table 7.3 gives examples of lenses achieved with a 60 µm thick layer. It was spun in 5 successive spinning steps [Völk95].

Table 7.3

lensØ/µm	f/µm	R/µm	h /µm	NA
2000	8770	4390	114	0.11
1400	4380	2190	115	0.16
1000	2270	1140	116	0.22
800	1540	770	112	0.25
600	900	450	115	0.32

7.1.2. Modelling of the melting process

The shape of the reflow lenses can be described using a polynomial approach with simple assumptions for the boundary conditions [Sher94]. Due to rotational symmetry a polynomial with only even exponents in x and y is needed. The profile of a lens can be approximated by:

$$z(\rho) = \sum_{n=0}^{\infty} a_{2n} \rho^{2n} \tag{7.5}$$

Melting of Photoresist

R is the lens radius and it is

$$R\rho = \sqrt{x^2 + y^2} \quad \text{with} \quad 0 \leq |\rho| \leq 1 \tag{7.6}$$

The following boundary conditions are valid

1) $\quad z(\rho)\,\big|_{\rho=1} = 0$

2) $\quad \dfrac{\delta z}{\delta \rho}\,\big|_{\rho=1} = -R \tan \alpha = A$

where α is the rim angle.

From the assumption in equation 7.3 it follows

$$V_{lens} = 2\pi R^2 T = 2\pi R^2 \int_0^1 \sum_{n=0}^{\infty} a_{2n}\, \rho^{2n+1}\, \delta\rho = \pi R^2 \sum_{n=0}^{\infty} \frac{a_{2n}}{n+1} \tag{7.9}$$

If the lens sag h is available as post melt information a six order approach for the polynomial is possible:

$$z(\rho) = a_0 + a_2\, \rho^2 + a_4\, \rho^4 + a_6\, \rho^6 \tag{7.10}$$

The equations for the coefficients a_0 through a_6 are

$$a_0 = h$$
$$a_0 + a_2 + a_4 + a_6 = 0$$
$$2a_2 + 4a_4 + 6a_6 = A \tag{7.11}$$
$$a_0 + \frac{a_2}{2} + \frac{a_4}{3} + \frac{a_6}{4} = T$$

with the solutions

$$a_0 = h$$
$$a_2 = -6h + 12T + \frac{A}{2}$$
$$a_4 = 9h - 24T - \frac{3}{2}A \tag{7.12}$$
$$a_6 = -4h + 12T + A$$

Fig. 7.4 – Typical lens shapes after melting

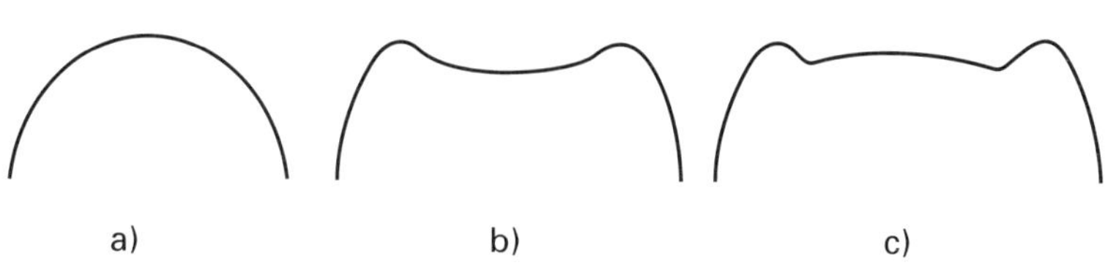

Fig. 7.4. shows typical lens shapes that occur after the melting process. A central bump as in fig. 7.4 c is also included in this mathematical treatment.

7.1.3. Postprocessing: Etching into substrate

The optical performance of photoresist lenses is excellent. However, the physical properties of the material have severe limitations. Photoresist is limited in its transmission, it is very sensitive to temperature and humidity, it is difficult to coat with antireflection layers and is troublesome to clean. Replication into another plastic material is one solution to overcome some of these problems. Another is etching the prepatterned photoresist together with the underlying substrate. This results in a transfer of the resist surface structure into the substrate (see fig. 7.1). Two methodes for eroding of the substrate material are widely used, ion milling [Grat93] and reactive ion etching (RIE) [Ster94] (see also chapter 2).

7.2. Applications

Combination with gratings

A combination of a refractive microlens with a deflection grating is a very attractive compact and small system. The deflection grating is easy to align behind the microlens as long as it fills the aperture of the lens. Two different integration methods have been reported [Saue94, Zhou95].

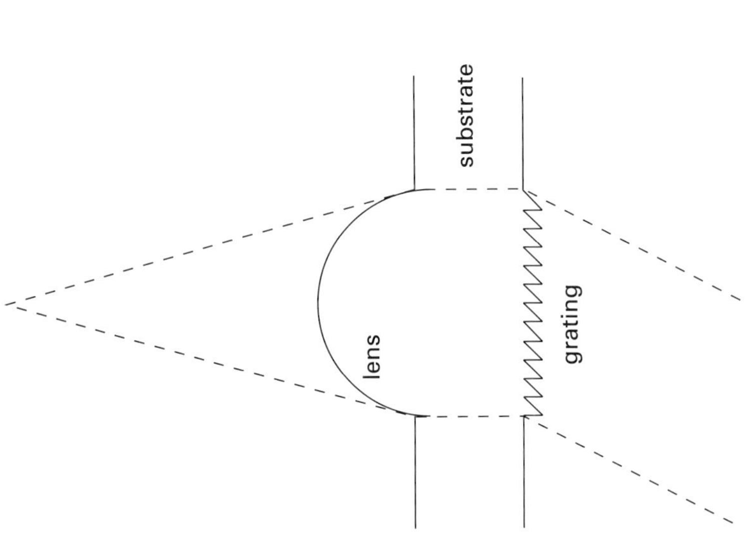

Fig. 7.5 – Microlens and grating etched into a substrate at opposite sides (After [Saue94])

The first method uses two opposite sides of a substrate (fig. 7.5) [Saue94]. The microlens is fabricated by the melting technique and is subsequently etched (RIE) into the substrate. The grating is etched on the opposite side. This method is very robust since only one substrate is necessary.

The second method situates both elements on the same side of the substrate [Zhou95]. This means that the grating is encapsulated by the lens. The grating must therefore have a higher index which is the case eg. for a Si_3N_4 film with n = 2.0. On top of the grating a PMMA film is spun with n = 1.5. Using PMMA as resist for the melting procedure is a little bit more complicated as a usual resist. Since the film thickness during one spinning cicle was only 2 μm, film deposition must be repeated. The maximum thickness practicable for UV - exposure should not exceed 9 μm. Melting can be done on a hot plate (300° C) for some minutes.

Deflection systems with pairs of microlenses

Deflection in a purely refractive way can also be done with lenses only. The principle is shown in fig. 7.6. [Hutl92].

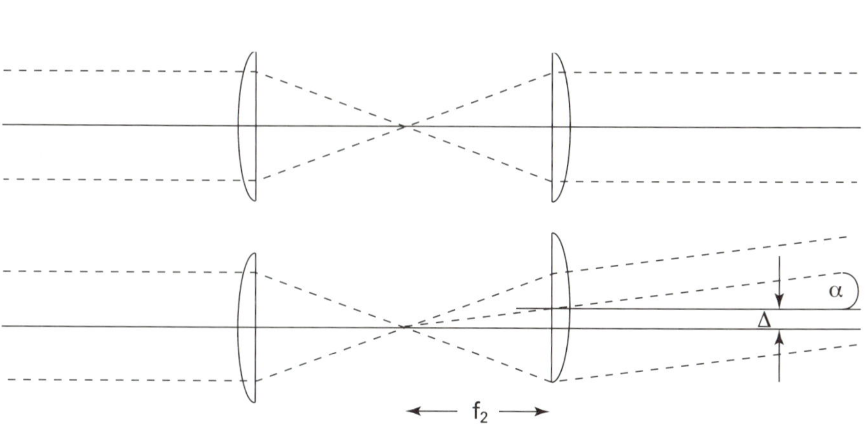

Fig. 7.6 – Use of an afocal pair of microlenses to deviate a beam of light
(After [Hutl92])

A pair of lenses is placed in a distance of the sum of their focal lenghts. If the lateral displacement of the lenses from the optical axis is zero, the displacement of the light after the second lens is also zero. However, if the second lens is displaced sideways, the ouput beam will be deviated by an angle which depends on the magnitude of the displacement and the focal length of the lens. The displacement α is given by

$$\alpha= \arctan \frac{\Delta}{f_2} \qquad (7.13)$$

If the substrate thickness can be chosen as the sum of both focal lengths this kind of deflection system can even be fabricated on one single substrate. By etching the resist lenses it is also possible to fabricate a very robust system. Alignment is much more critical as in the case above (see also chapter 2). But the efficiency of a purely refractive approach is higher.

This idea has found application in a hybrid opto-electronical interconnection system [Hend95]. It is used for a perfect shuffle interconnect where light deflection is performed with microlenses. This system has several shuffle stages. Those with short interconnects are realized electronically. But long interconnects with larger deflection angles make use of laterally displaced microlenses.

Also a piezo-driven dynamical version of light deflection by microlenses has been reported for use as a beam scanner [Mota94]. In this case two diffractive microlens arrays, one with positive lenses and the other with negative lenses, can be moved by piezo-electric transducers. The microlens arrays have been etched in fused quartz. The array size has been 10 mm on a side and 0.4 mm thick. This yields a total weight of each array of 140 mg. The focal length of the positive array has been 1200 μm and that of the negative array has been 1195 μm. The microlenses have been arranged in a 200 μm x 200 μm square. With compact actuators having a maximum translation range of ±25 μm it has been possible to achieve a 16 x 16 spot pattern with a scanner rate of 300Hz.

Combination with liquid crystals

If microlenses are immersed in liquid crystals then their focal length can be expressed as

$$f = \frac{R}{n_{LC} - n_{RE}} \qquad (7.14)$$

In the reported experiment the uniaxial birefringent liquid crystal material has an extraordinary refractive index, n_{LC}, between 1.74 and the ordinary index of 1.52 [Comm95]. The liquid crystal can be reorientated by an electric field so that the extraordinary refractive index is modulated between these two values. The refractive index n_{RE} of the photoresist was 1.64. Thus the modulated lens can act as a positive or negative lens. Lens diameters in the range between 50 μm and 200 μm have been realized. Lenses have been built with variable focal length from -910 μm, with no voltage, to between 380 μm and 560 μm.

Etching a resonator cavity

Experimental results of ion milling into YAG have been reported in [Tara95]. The application is illustrated in fig. 7.7.

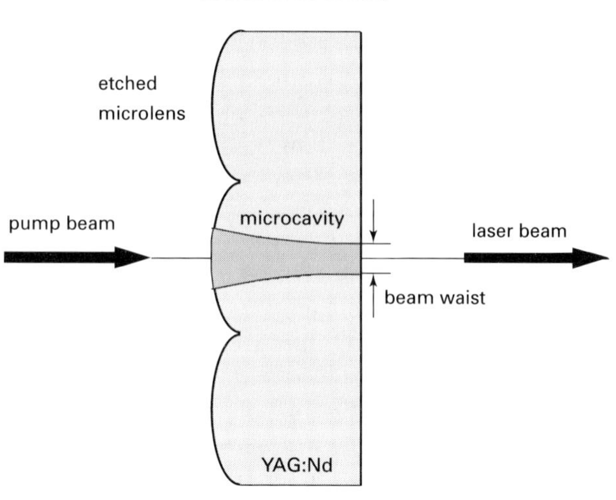

Fig. 7.7 – Use of microlenses etched in YAG as resonator cavity
(After [Tara95])

One side of the substrate is etched with spherical shape to form a micro-cavity. Selectivity, defined as the ratio between the etching rates of the substrate and the resist mask, can be adjusted in a range between 0.3 and 3 by controlling the oxygen gas flow. A typical transfer from the former resist lens to the YAG lens can be seen in fig. 7.8.

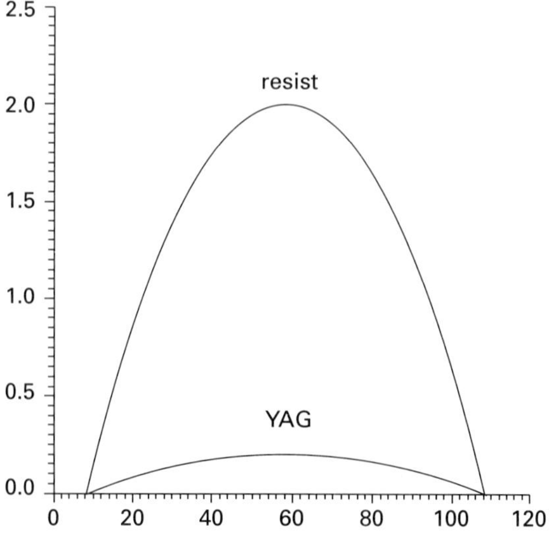

Fig. 7.8 – Typical profile of microlens before and after etching [Tara95]

References for chapter 7

[Artz92]   Guy Artzner: "Microlens arrays for Shack-Hartmann wavefront sensors", Optical Engineering, p. 1311 - 1322, 1992.

[Comm95] L. G. Commander, S. E. Day, C. H. Chia, D. R. Selviah: "Microlenses immersed in nematic liquid crystal with electrically controllable focal length", in Microlens Arrays, Vol. 5, 1995 EOS Topical Meetings Digest Series, pp. 72 - 76.

[Daly90]   D. Daly, R. F. Stevens, M. C. Hutley, N. Davies: "The manufacture of microlenses by melting photoresist", J. MEAS. Sci. Technol. 1, P. 759 -766, 1990.

[Gex96]   F. Gex, D. Horville, G. Lelièvre, and D. Mercier: "Improvement of a manufacturing technique for long focal length microlens arrays", to appear in Pure and Applied Optics, 1996.

[Grat93]   E. J. Gratix: "Evolution of a microlens surface under etching conditions", in Miniature and Micro-Optics and Micromechanics, N. C. Gallagher, Jr., C. Roychoudhuri, ed., Proc. SPIE 1992, p. 266 - 274, 1993.

[Hend95]   W. L. Hendrick, Ph. J. Marchand, F. B. McCormick, I. Çokgör, and S. C. Esener: "Optical Transpose Interconnection System: System Design and Component Development", in Optical Computing, Vol. 10, 1995 OSA Technical Digest Series, pp. 283-285.

[Hutl92]   M. C. Hutley, P. Savander, and M. Schrader: "The use of microlenses for making spatially variant optical interconnections", Pure and Applied Optics, Vol. 1, p. 337 - 342, 1992.

[Jay94]   T. R. Jay and M. B. Stern: "Preshaping photoresist for refractive microlens fabrication", Optical Engineering, Vol. 33, No. 11, p. 3552 - 3555, 1994.

[Mota94]   M. E. Motamedi, A. P. Andrews, W. J. Gunning, and M. Khoshnevisan: "Miniaturized micro-optical scanner", Optical Engineering, Vol. 33, No. 11, p. 3616 - 3623, 1994.

[Saue94]   F. Sauer, J. Jahns, C. R. Nijander, A. Y. Feldblum, and W. P. Townsend: "Refractive-diffractive micro-optics for permutation interconnects", Optical Engineering, Vol. 33, No. 5, p. 1550 - 1560, 1994.

[Sher 94]   J. T. Sheridan, J. Schwider, N. Streibl, S. Haselbeck, M. Eisner, M. Heissmeier, and O. Falkenstörfer: "Modelling and measurement of melted microlens shapes", Optik, Vol. 97, No. 4, p. 174 - 180, 1994.

[Ster94]   M. B. Stern and T. R. Jay: "Dry etching for coherent refractive microlens arrays", Optical Engineering, Vol. 33, No. 11, p. 3547 - 3551, 1994.

[Tara95]   V. Tarazona, M. Rabarot, E. Molva, V. Marty, J. M. Combes: "Long focal length microlenses in laser materials for the stabilization of microchip lasers", in Microlens Arrays, Vol. 5, 1995 EOS Topical Meetings Digest Series, postdeadline paper P3.

[Völk95]   R. Völkel, P. Nussbaum, K. J. Weible, H. P. Herzig, R. Dändliker, S. Haselbeck, M. Eisner, and J. Schwider: "Fabrication of non-conventional microlens arrays", in Microlens Arrays, Vol. 5, 1995 EOS Topical Meetings Digest Series, p. 116 - 120, 1995.

[Zhou95]   Z. Zhou and T. Drabik: "Coplanar refractive-diffractive doublets for optoelectronic integrated systems", Applied Optics Vol. 34, No. 17, p. 3048 - 3054, 1995.

Excimer Laser for Micro-optics

More than 20 years ago excimer laser entered laboratories for use in optical spectroscopy. Since 1980 they have found new applications in micro-electronic circuit fabrication. Their characteristic is short wavelength in the UV region combined with high optical power in the order of MW/cm^2. Shorter wavelengths are of great interest for lithography. The most commonly known application of excimer lasers is photoablation. Furthermore chemical effects appearing below the ablation threshold can find applications in micro-optics.

8.1 Photoablation

8.1.1. Basics of the process

Photoablation occurs if an irradiated material absorbes energy above a certain threshold which leads to a spontaneous vaporization (fig. 8.1.).

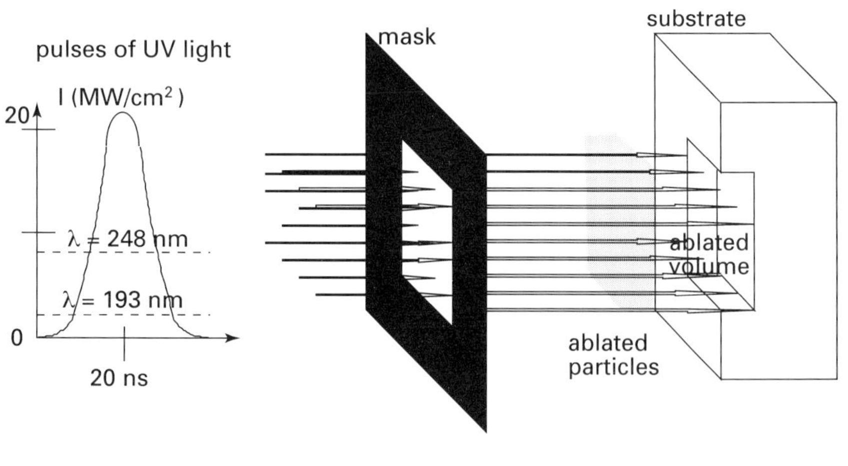

Fig. 8.1 – Illustration of photoablation. Intensity for one laser pulse is given with the ablation thresholds for λ = 193 nm and λ = 248 nm (After [Laza90]).

Chemical chain scission and volume heating occur at the same time. The most important parameter in this process is absorption. Equation 8.1 is the Fresnel law for reflectivity including absorption which in general is the imaginary part of the refractive index of a given material.

$$R = \frac{(n-1)^2 + k^2}{(n-1)^2 + k^2} \qquad (8.1)$$

k is related to the absorption coefficient α

$$k = \alpha \frac{\lambda}{4\pi} \qquad (8.2)$$

α is given in units of cm^{-1}. Often one can find the inverse parameter p = $1/\alpha$ in units of Å. p denotes the depth where the absorbed light intensity is 65 %. Light absorption is controlled by the Beer-Lambert law:

$$I(x) = I_0 (1-R) e^{-\alpha x} \qquad (8.3)$$

It should be noted that constant absorption coefficients are not strictly valid under ablation conditions.

Table 8.1 lists several polymer material and their absorption parameters. The polymers are PET (polyethylene terephtalate), PPQ (polyphenylquinoxaline), PS (polystyrene), and PMMA (polymethyl methacrylate) [Laza90].

Table 8.1

polymer	193 nm		248 nm	
	$\alpha(cm^{-1})$	p(Å)	$\alpha(cm^{-1})$	p(Å)
PET	300	330	160	625
PPQ	65	1500	37	2700
PS	800	125	6.1	16400
PMMA	2	50000	0.065	1500000

The ablated volume does not only depend on the energy given by the wavelength but also on the pulse duration and the material ablation threshold. For the same polymers and wavelengths as before table 8.2 shows the ablation treshold [Laza90].

Table 8.2

polymer	ablation threshold for 193 nm		ablation threshold for 248 nm	
	(MW/cm^2)	(mJ/cm^2)	(MW/cm^2)	(mJ/cm^2)
PET	1.4	20	1.96	28
PPQ	1.26	18	2.24	32
PS	0.77	11	2.8	40
PMMA	-	-	6.8	100

8.1.2. Applications of photoablation

During irradiation a certain volume is ablated in the very short time of one laser pulse. However the explosive ablation also roughens the surfaces as can be seen in the following photos. Fig. 8.2 illustrates the bottom roughness of holes fabricated by excimer laser ablation with different wavelengths. The shortest wavelength shows the smoothest surface [Laza90]. Additionally problems occur related to heating of the material during irradiation which can lead to locally melted areas. Therefore the side walls are not optically clean enough to act as a micro-prisms or -mirrors for beam deflection.

Fig. 8.2 – Bottom of ablated holes using different wavelength:
a) KrF 248 nm, b) XeF 351 nm, c) ArF 193 nm
(Courtesy Sylvain Lazare, [Laza90])

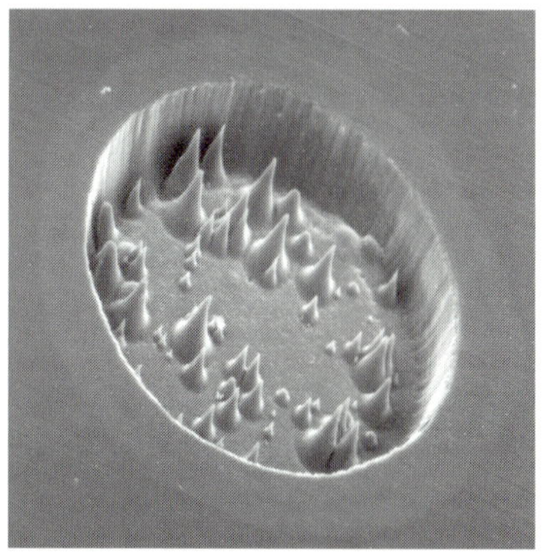

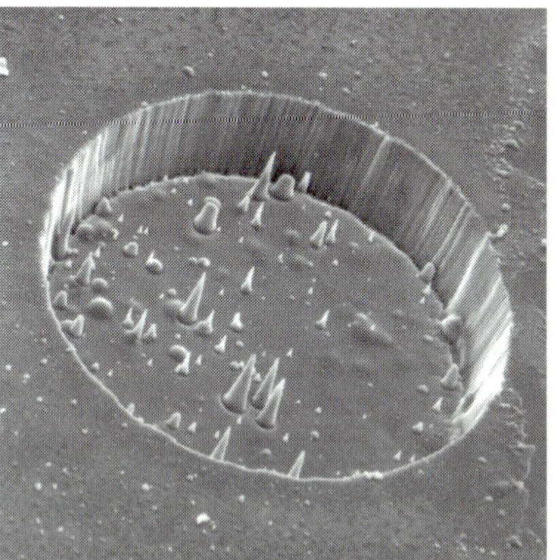

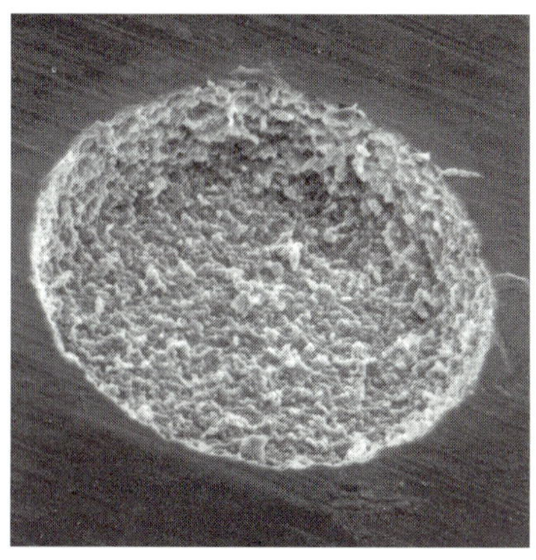

Fiber holder arrays

Besides the question of surface roughness the ablation process is suitable for fabricating holder structures eg. fiber holders. One-dimensional and two-dimensional holders have been reported:

- One-dimensional fiber holders

As already discussed in chapter 4, V-grooves are used for the connection of waveguides to optical fibers. But if waveguides are fabricated in plastic material it may sometimes be convenient for the connector to also be a plastic. Then a monlithic approach for waveguide and holder is possible in order to reduce misalignment. Such an approach has been reported in [Boot92].

Fig. 8.3 – Polyguide ablated structures for coupling to waveguides
(After [Boot92])

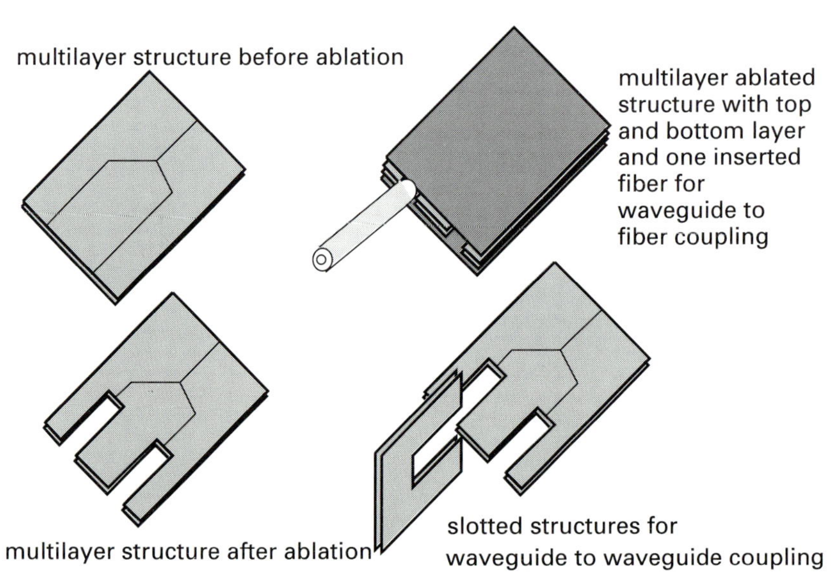

The waveguide material is Polyguide (du Pont de Nemour). It is choosen because it is suitable for both light guiding and laser ablation. These waveguides show optical losses in the region of 0.1 dB (830 nm), 0.2 dB (1300 nm) and 1.2 dB (1550 nm). Typical ablation conditions are a wavelength of 248 nm, a repetition rate of 10 Hz and a dose of 1 - 2 J/cm^2. For fiber insertion small rectangular slots with a width of 115 - 123 µm and a length of about 600 µm are fabricated (fig. 8.3). The sidewall inclination is about 4°. Because Polyguide is a flexible material the slot can have slighlty smaller width than the fiber diameter of 125 µm in order to self-center the fiber in the elastic slot. In a sandwich structure shown also in fig. 8.3 the self-centering occurs in vertical direction, too. Fibers without the plastic coating and slightly bevelled are brought into the slots together with UV curing glue. Good coupling efficiency has been reported for fiber waveguide coupling which is better than 0.5 dB. With this technique waveguide to waveguide coupling and fiber to fiber coupling can also be achieved (fig. 8.3).

- Two-dimensional fiber holders

Two-dimensional arrays have been fabricated in polyimide with a thickness of 1 mm [Prou94]. The laser wavelength was 308 nm and the incident fluence was about 10 to 14 J/cm^2. The holes were machined serially so that the pitch accuracy between two neighbouring holes was given by the accuracy of the translation stages. An error in the ablated hole diameter was determined to be ±3 µm. The entrance hole diameter was different from the observed exit hole diameter. Due to the anisotropic intensity distribution of the excimer beam the entrance hole was shaped elliptically with a diameter of 200 µm and 160 µm, respectively. The exit holes were circular with a diameter of 125 µm. The taper angle between entrance and exit was about 2°. The absolute alignment accuracy was about ± 5 µm and not sufficient for single-mode fiber coupling. These high tolerances occured because the ablated holes could not be adapted to the fiber diameter. A solution for that might be to use another material, eg. Polyguide. The flexibility of the material allows the fabrication of holes with slightly smaller diameter. Improvements have also been achieved using additional alignment structures [Sasi94]. In this case sophisticated alignment effort is required.

Lens fabrication by ablation and melting

A lens fabrication method using photoablation combined with the melting technique already discussed in chapter 7 is reported in [Miha93]. Teflon AF is chosen for a suitable substrate because of its high transparency over a wide range of wavelengths (190 nm - 2000 nm). Teflon AF is an amorphous form of the almost opaque Teflon and is manufactured by Du Pont. Since the substrate is also transparent for the ablation wavelength of 193 nm it is made artificially absorbant by doping it with tris(perfluoroalkyl)-s-triazine. For thin film preparation Teflon AF and the absorber material are mixed and spin coated onto a silicon substrate. Film thicknesses of the order of 100 µm can be achieved. For ablation a mask is projected onto the substrate with a variable imaging factor. With this method different lens sizes can be fabricated using only one mask. The ablation process preshapes the islands to be melted. The substrate is then heated to 160°C to remove the dopant. After the dedoping process, the polymer is heated to its melting temperature of 300°C for a period of 20 minutes.

8.2. Monomer diffusion

8.2.1. Basics of the process

Besides the photoablation technique the eximer laser can also be used below the ablation threshold in order to create radiation damage in PMMA [Laza96]. As already explained in chapter 2, fig. 2.2, PMMA suffers main chain scission when irradiated with deep UV light. Linear PMMA with high molecular weight shows a reduction of molecular weight in UV irradiated regions. In contact with monomer vapour, the molecules from the vapour diffuse into the irradiated domains and cause a significant volume expansion which can be stabilized by UV polymerization (this swelling process will be discussed in more detail in chapter 10). Fig. 8.4 shows the principle of the process.

Fig. 8.4 – Principle of fabrication process: a) KrF Excimer laser (248 nm) irradiation of PMMA, styrene diffusion, c) Mercury lamp (365 nm) copolymerization (After [Laza96])

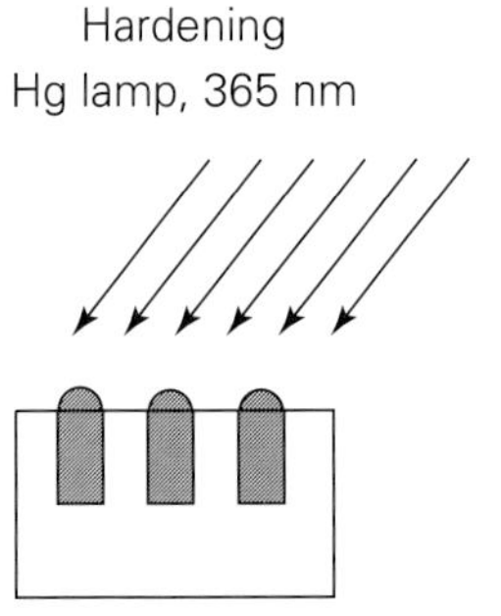
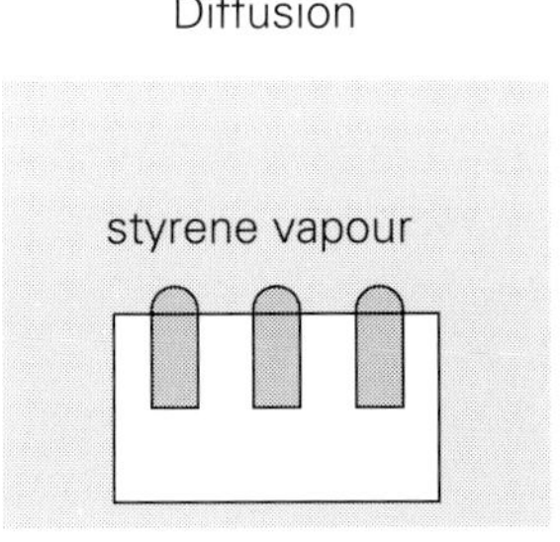
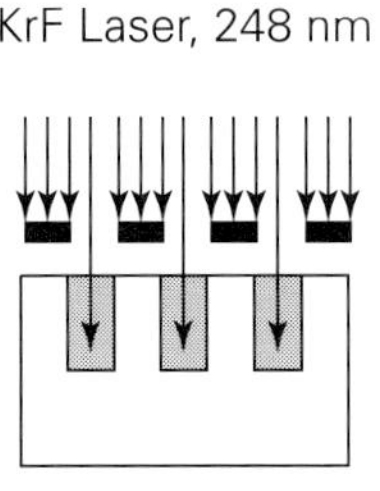

8.2.2. Applications of monomer diffusion

Lens fabrication by monomer diffusion

For the fabrication of lenses a mask with circular apertures is used. In this case the mask can either be imaged onto the substrate or contact printed since the energy fluence is only about 10 mJ/cm^2 which is far away from the ablation threshold of 100 mJ/cm^2. After irradiation the sample is placed into a hot monomer vapour athmosphere typically with a temperature of 90°C and a diffusion time of 45 minutes. The diffusion of monomer vapour causes the irradiated domains to swell, and by surface tension lenslike shapes appear on top of the irradiated areas. In order to prevent outdiffusion the sample is irradiated again with UV light (λ = 365 nm). A fluence of about 10 mW/cm^2 initiates a copolymerization process which hardens the lenses [Laza96].

Because of the exponential decrease of UV light in PMMA and the deep wavelength the changes in molecular weight are significant for swelling only until a certain depth. This depth can be measured eg. with confocal micro Raman spectroscopy. The spectroscopy has detected a new bond vibration caused by irradiation up to a depth of about 110 μm. This is at the same time the limit for styrene diffusion in the irradiated domains and determines the potenial swelling volume. Fig. 8.5 shows saturation at higher doses. A typical lens array is shown in fig. 8.6.

Fig. 8.5 – Relative height of microlenses with a diameter of 200 mm versus a dose at 248 nm, 10 mJ/cm2 and 20 Hz, (100 % = 10.6 mm)
(After [Laza96])

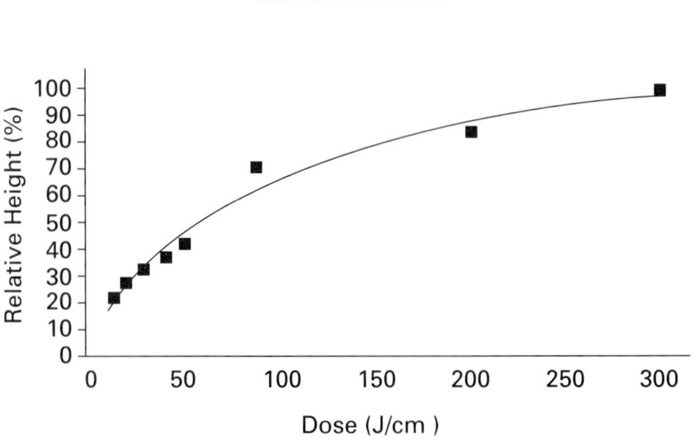

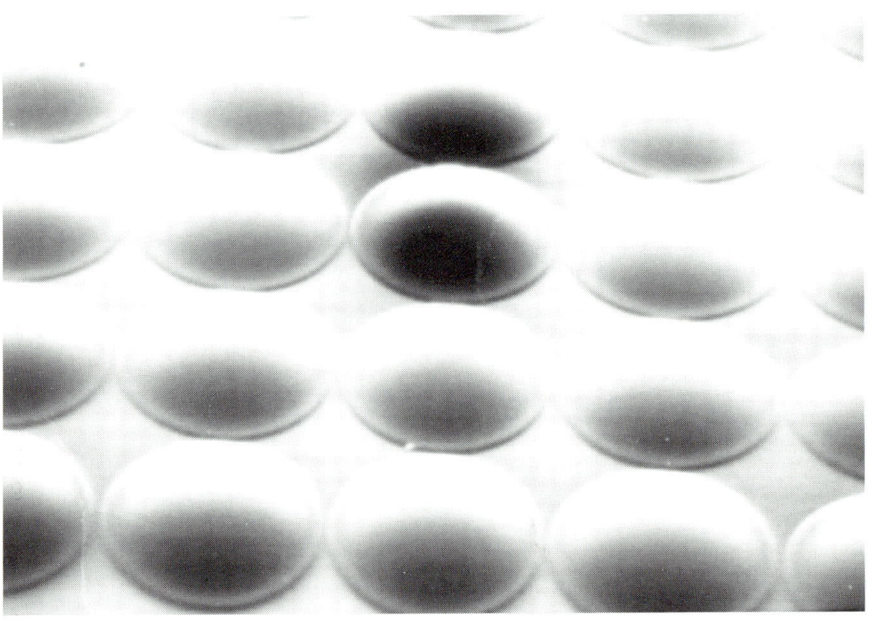

Fig. 8.6 – Microlensarray fabricated by excimer irradiation and monomer diffusion. Lens diameter is 200 mm [Laza96]

Prism fabrication by monomer diffusion

With a similar process as already described for lens fabrication in the previous section it is possible to swell microprisms [Rohr95]. The process also uses the potential of swelling PMMA according to different levels of degradation. In this case especially the high power excimer laser can be replaced by a high pressure mercury lamp with a wavelength of 365 nm. The resist is a mixture of 90% MMA oligomer (methyl methacrylate) and 25% of ketal (2,2-dimetoxy-2-penylacetophenone). Ketal is a photosensitive substance that splits into radicals when illuminated with UV light. A 20 μm thick layer is spun on a glass substrate and baked for 1 hour at 40°C. Then this layer is irradiated via a grey scale mask using mask projection onto the substrate. The illumination parameters are chosen accordingly. During irradiation the sample is simultanously exposed to MMA vapour. Those domains with newly created radicals allow the MMA vapour to diffuse in, which causes volume expansion. The amount of volume expansion is controlled by the grey scale taking into account the dependancy beween radical creation and intensity. With this method especially linear profiles can be achieved which act as prisms. A typical example is shown schematically in fig. 8.7.

Fig. 8.7 – Schematic of a 3x3 prism array fabricated by monomer diffusion. The maximum phase shift achieved was 8π. (After [Rohr95])

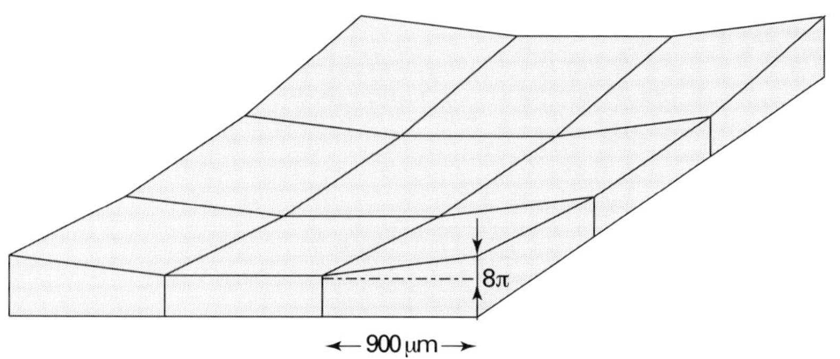

8.3. Index change via radiation

8.3.1. Basis of the process

Related to the degradation of polymers when they are exposed to UV irradiation is a change in the refractive index.

8.3.2. Application: Waveguide fabrication

Irradiation with a wavelength of 248 nm (excimer laser) or 256 nm (mercury lamp) cause a sufficient change in refractive index to create a waveguide with a depth of about 5 µm [Fran94]. In order to have a well defined waveguide region it is best to spin a 5 µm thick PMMA layer on top of a glass or silicon substrate. The change in refractive index is shown in fig. 8.8 when the substrate was irradiated with a mercury lamp (256 nm) from a mask aligner. A maximum index change of about 0.015 can be achieved after 1 hour of irradiation which corresponds to the saturation of the process for this material.

A variation of the process uses embossed structures (see also chapter 4.2) [Fran94]. A heated nickel mould is embossed in PMMA. Then the sample is irradiated globally by UV light without any mask. This converts the whole surface into a planar waveguide but only the waveguides on the bottom of depression are used as strip waveguides.

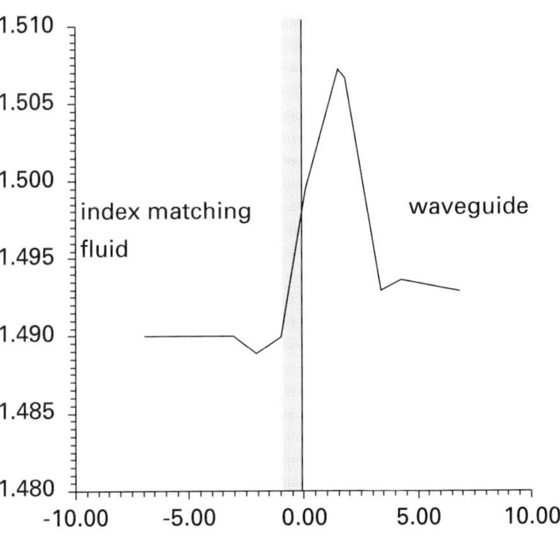

Fig. 8.8 – Index profile perpendicular to the surfaces of planar waveguides in bulk material after saturation (After [Fran94])

References for chapter 8

[Boot92]   B. L. Booth in 'Polymers for lightwave and integrated optics', chapter 9, "Optical interconnection polymers", p. 231 - 266, ed. L. A. Hornak, Marcel Decker Inc., New York 1992.

[Fran94]   W. F. X. Frank, B. Knödler, A. Schösser, T. K. STrempel, T. Tschudi, F. Linke, D. Muschert, A. Stelmmaszyk, H. Strack, A. Bräuer, P. Dannberg, and R. Göring: "Waveguides in Polymers", in 'Fiber Optic Materials and Components', Proc. SPIE 2290, p. 125 - 132, 1994.

[Laza90]   S. Lazare: "Photoablation par laser ultraviolet: gravure et modification de surface de polymeres", chapter 7, in 'Techniques d'utilisation des photons: Principe et Applications', Collection ELECTRA Editions DOPEE85, 1992.

[Laza96]   S. Lazare, J. Lopez, J.-M. Turlet, M. Kufner, S. Kufner, and P. Chavel: "Microlenses fabricated by UV excimer laser irradiation of PMMA followed by styrene diffusion", Applied Optics, Vol. 35, No. 22, p. 4471 - 4475, 1996.

[Miha93]   S. Mihailov and Sylvain Lazare: "Fabrication of refractive microlens arays by excimer laser ablation of amorphous Teflon", Applied Optics, Vol. 32, No. 31, p. 6211 - 6218, 1993.

[Prou94]   G. M. Proudly, C. Stace, and H. White: "Fabrication of two-dimensional fiber optic arrays for an optical crossbar switch", Optical Engineering, Vol. 33, No. 2, p. 627 - 635, 1994.

[Rohr95]   A. Rohrbach, K.-H. Brenner: "Surface-relief phase structures generated by light-initiated polymerization", Applied Optics, Vol. 34, No. 22, p. 4747 - 4754, 1995.

[Sasi94]   J. M. Sasian, R. A Novotny, M. G. Beckman, S. L. Walker, M. J. Wojcik, and S. J. Hinterlong: "Fabrication of fiber bundle arrays for free-space photonic switching systems", Optical Engineering, Vol. 33, No. 9, p. 2979 - 2985, 1994.

Deep Etch Lithography with Synchrotron Radiation - LIGA

The german acronym LIGA stands for Lithographie, Galvanoformung und Abformung. It represents a deep etch fabrication method which has been developed since the beginning of the 80's. Structures in different types of materials like plastics, metals or ceramics can be several hundreds of micron deep with very steep side walls. It is a process suitable for cheap mass fabrication using molding techniques.

9.1 Basics of the process

9.1.1. The LIGA process

LIGA has been developed for the fabrication of microstructures with high structural depth and high aspect ratio [Beck82, Beck86, Ehrf88]. It is a combination of lithography using synchrotron irradiation, galvanoforming and moulding. Fig. 9.1 shows the principle of the LIGA process [Ehrf87]. Using a special gold mask (fig. 9.2) a thick layer of PMMA is irradiated [Münc87,Heub88]. The PMMA is deposited on a metal plate which is useded later for galvanoforming. For sufficient dose deposition in more than 100 μm depth a highly intense and parallel source with a quantum energy of several keV is needed. Therefore synchrotron irradiation is most suitable. During irradiation the same chemical reaction takes place as for a thin layer of positive photoresist (fig. 2.3.). With a special developer it is possible to dissolve only the irradiated regions [Ghic81]. The anisotropic etch rate is a factor 1000 higher in the irradiated regions compared to the original material [Mohr88]. After development the galvanoformig process follows. The metal base plate serves as an electrode. Metal is deposited in the developed volume (eg. nickel or copper). The galvanoprocess is continued until the empty volume is filled and additionally an opposite thick base plate is formed. Then the original base plate and the remaining PMMA is removed by etching. This yields an inverse master for a subsequent moulding process. Replication can easily be done in plastics and ceramics for a high number of replicas.

9.1.2. Mask preparation

For the lithographic process special masks have to be prepared, which are either fully transparent or fully opaque to the synchrotron radiation. This means that requirements for thickness and aspect ratio are much higher than for other masks that have been described so far in chapter 2. To achieve a sufficient contrast between irradiated and nonirradiated regions the mask has to have a certain thickness in the order of 10 μm to 20 μm. This height is machined in successive irradiation and galvanoforming processes. Fig. 9.2 shows the principle for the fabrication. The mask is grown on a titanium foil . On a very smooth Invar block a titanium layer with a thickness of about 2 μm is sputtered. After cooling the titanium layer has the required strength which is necessary to support the gold mask and to provide a sufficiently flat surface. A large hole of about 20 x 60 mm^2 is etched in the Invar block which is stopped by the titanium. The result is a free 2 μm thick titanium layer. Next a 3 μm to 4 μm thick PMMA layer is spun onto the titanium. An electron beam writer with high spot resolution patterns the PMMA. After development the resist is filled by glavanoforming with a 2 μm to 3 μm thick gold structure. Here the titanium foil is used as an electrode.

Fig. 9.1 – Fabrication steps for the LIGA process (After [Ehrf87])

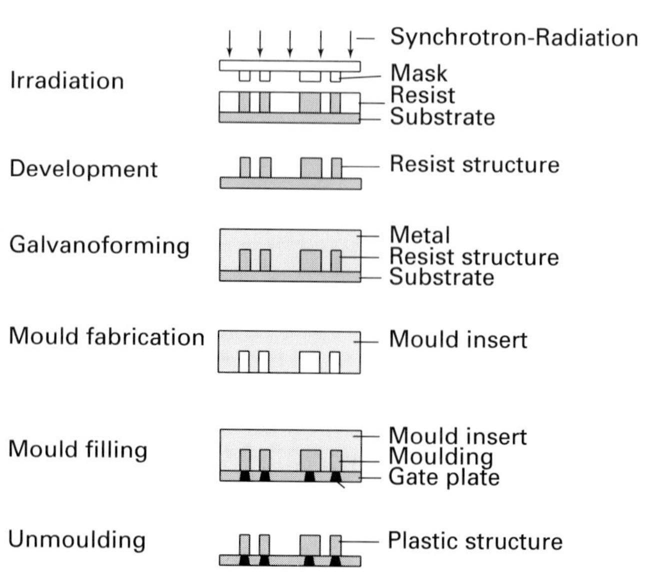

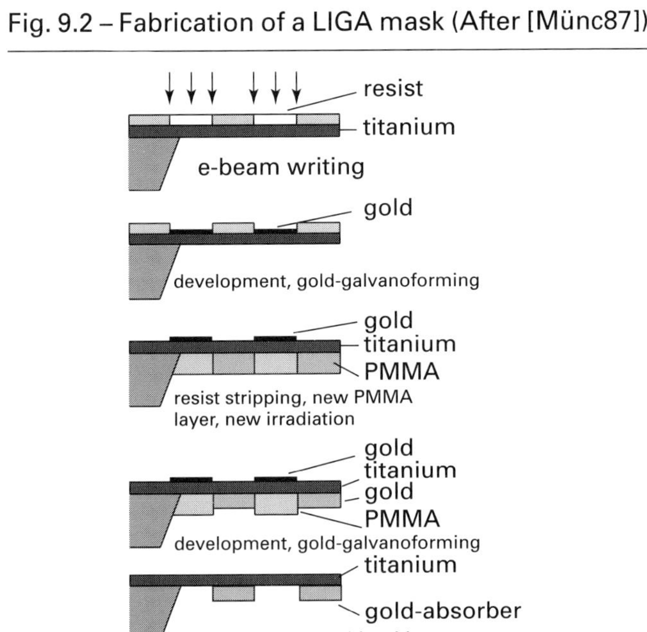

Fig. 9.2 – Fabrication of a LIGA mask (After [Münc87])

To protect the other side of the foil it is covered with a 20 µm thick PMMA layer which represents the next resist layer to be structered. After the thin PMMA layer has been stripped the gold mask can be used to pattern the opposite 20 µm thick PMMA by synchrotron irradiation. After development the PMMA is filled with another gold structure with a height of about 10 µm to 20 µm. This gold mask is the final mask for the deep etch process described above in fig. 9.1.

9.2. Fabrication of refractive micro-optical elements

9.2.1. Microprisms

The LIGA process allows the fabrication of microstructures with high structural depth and steep side walls on a base plate (LIGA without galvanoforming and moulding). These steep side walls act as microprisms or micromirrors [Bren93, Schi94]. In this case the optical quality of the surfaces is important. An example of an interferometric measurement is shown in fig. 9.3 representing a typical LIGA side wall with a height of 500 µm. The interferogram shows a very smooth optical surface over a lateral extension of 500 µm and a roughness of better than 50 nm. The top and bottom of the structure the dose deposition show small grooves, but more than 90 % of the area is extremely steep. The lateral deviation from an ideal vertical wall is about 100 nm per 100 μm depth. Also visible in the interferogram is the way in which each imperfection of the mask is projected into the structure in form of a groove over the entire structural depth, which motivates the high effort for mask preparation.

Fig. 9.3 – Interferometric measurement of a LIGA side wall with a height of 500 μm (After [Brenn93])

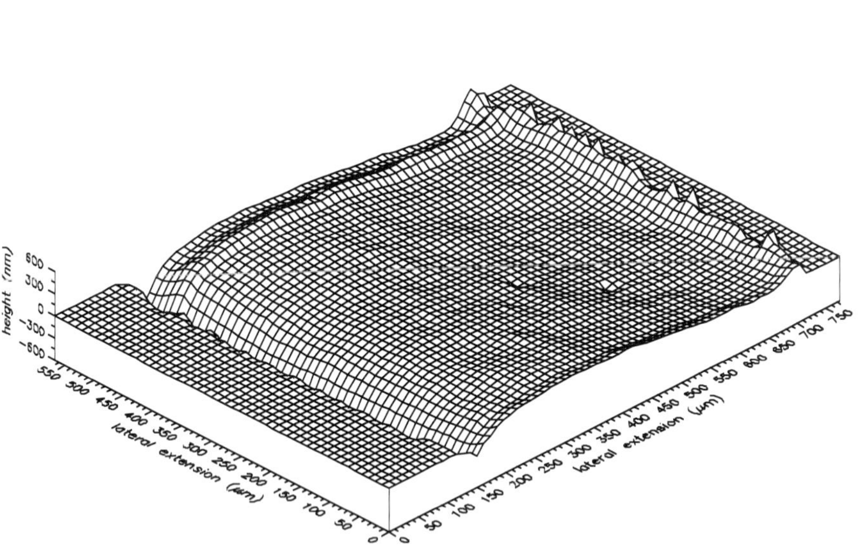

9.2.2. Microbeamsplitter

The precision of the LIGA process allows the fabrication of miniaturized beamsplitters in analogy to conventional beamsplitters [Kufn93].

This principle is illustrated in fig. 9.4. Two different substrates contain the two building blocks of the beamsplitter. The hypothenuse of one of the building blocks is coated with a dielectric layer. Then the two building blocks are stacked together as indicated. Alignment structures ensure correct positioning. Additional optical glue fixes the beamsplitter in its final position. A photograph of such a beam splitter can be seen in fig. 9.5. With much less effort aperture beamsplitting is also possible.

Deep Etch Lithography with Synchrotron Radiation - LIGA

Fig. 9.4 – Principle of a refractive beamsplitter fabricated with LIGA
(After [Kufn93])

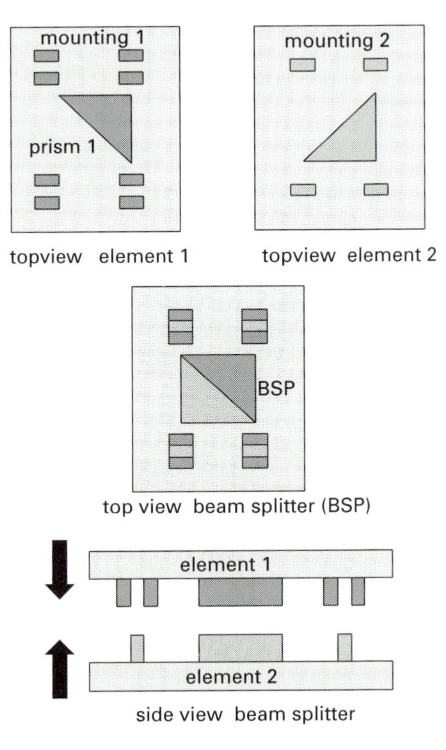

Fig. 9.5 – Photograph of a LIGA beamsplitter consisting of two parts on different substrates (testsample courtesy of IMT, Forschungszentrum Karlsruhe, [Kufn93])

9.2.3. Microlenses

Cylindrical microlenses fabricated directly by LIGA

Cylindrical microlenses can be achieved by transferring curved structures from the e-beam mask to the substrate. With the lithographic precision of the process the designed curvature and shape and thus the focal length of a cylindrical microlens can be achieved with high precision. Since there is no design freedom in depth (the mask is directly projected) the direct process is restricted to cylindrical lenses. Fig. 9.6. shows dimensions of a typical cylindrical lens and the measured focal beam characteristics [Gött92].

In order to overcome the missing vertical curvature an optical fiber can be placed behind a cylindrical microlens [Schi94]. But this solution is restricted by the fixed curvature of the fiber. Another possibility is to use crossed cylindrical microlenses [Schi94]. In this case two substrates have to be aligned relative to each other. This can be supported with holder structures as eg. in the case of a LIGA beamsplitter.

Fig. 9.6 – Measurement of focal distance of a cylindrical microlens
(After [Gött92])

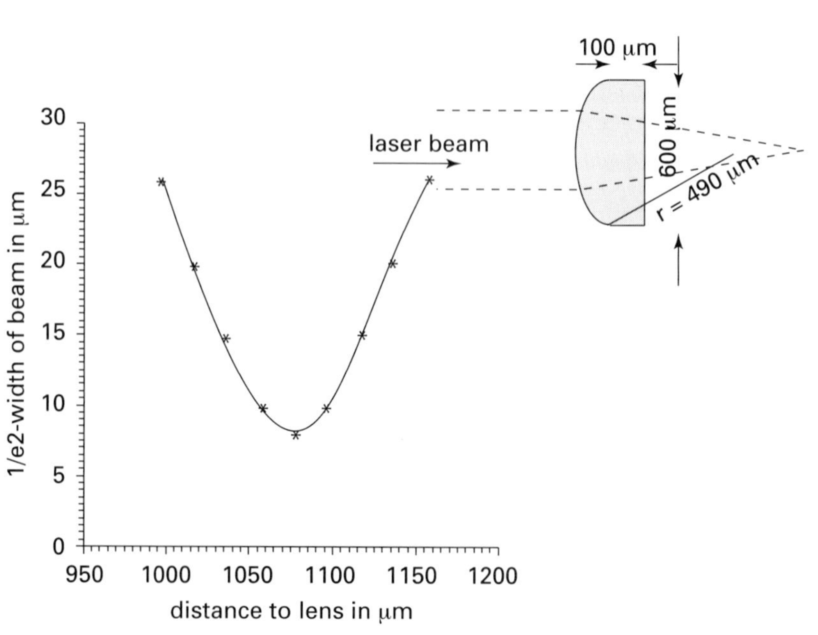

Microlenses fabricated by monomer diffusion

The same microlens fabrication process as already described in chapter 8 using UV irradiation and subsequent monomer diffusion also works if laser light is replaced by synchrotron light [Frit94]. If the substrate is not placed on a metal base plate, lenses can be swollen on both sides of the substrate. Since the intensity distribution of electromagnetic radiation decreases exponentially into a substrate, the top side is irradiated with higher dose compared to the bottom side. This causes an asymmetry between the swelling behaviour of the top and the bottom side.

Microlenses fabricated by melting

In chapter 7 the fabrication of microlenses with melting of photoresist has been discussed. That process is restricted by the limits of layer thickness from spin coating. But synchrotron irradiation can structure PMMA much deeper. This potential of depth can be used for the fabrication of spherical microlenses as reported in [Gött95a]. For this application crosslinked PMMA is used as initial substrate material. The layer thickness can be varied between 50 μm and 500 μm. Lenses were fabricated in a diameter range between 10 μm and 1000 μm. The fabrication procedure is sketched in fig. 9.7. With the LIGA process cylindrical islands are created in the PMMA. After development these islands have to be irradiated again without any mask in order to partly crack the polymer network. Knowing the absorbed dose over the resist depth a well defined molecular weight distribution in the islands is achieved. A temperature slightly above the glass temperature of PMMA transfers the islands into hemispherical lenses or depending on the height of the cylinder into a sphere segment.

Micro-optics and Lithography

Fig. 9.7 – Fabrication of microlenses by melting LIGA structures
(After [Gött95a])

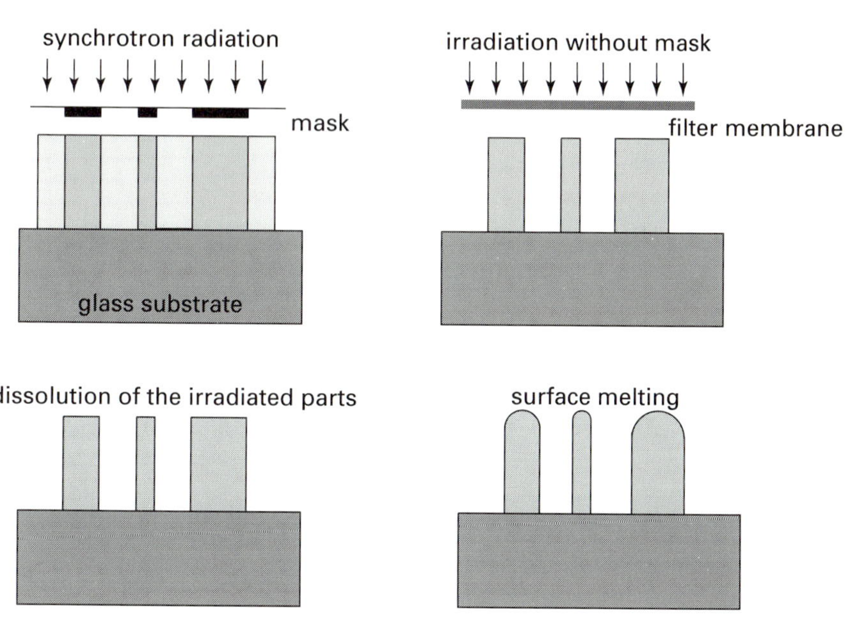

Fig. 9.8 – Hemisphere of app. 250 mm diameter with remaining
cylinder; total height of the structure is app. 230 mm
(Courtesy J. Göttert, Forschungszentrum Karlruhe, [Gött95a])

9.3. Applications in optical microsystems

9.3.1. 2½ D architecture

The possibility to fabricate structures with high structural depths and steep side walls offers a new architectural principle for the arrangement of micro-optical elements which is different from the guided wave approach (chapter 3) and also from the idea of stacked optical systems (chapter 6). The structures can be precisely placed in two dimensions on a base plate in a plane. This base plate can be used like an optical table in macro-optical setups. In addition the structural height can be used for free space optical imaging. "½" means that the height is limited to about several hundred micron (<1 mm). The architectural concept can be seen in fig. 9.9 [Kufn93].

Fig. 9.9 – Principle of a 2½ D architecture (After [Kufn93])

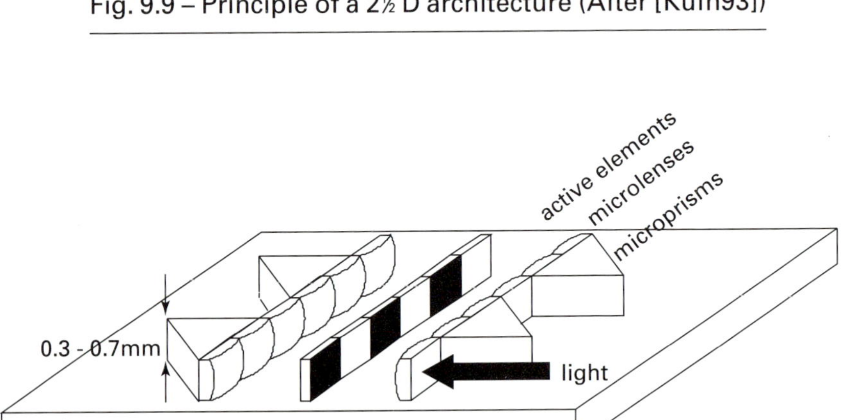

On a substrate (base plate) mounting elements and deflecting prisms are situated. In addition lines of microlenses and active elements can be aligned relative to each other with the help of holder structures. Such a typical LIGA layout can be seen in fig. 9.10.

Fig. 9.10 – Layout of 2½ D LIGA structure with lines of prisms and holder elements (testsample courtesy of IMT, Forschungszentrum Karlsruhe, [Bren93])

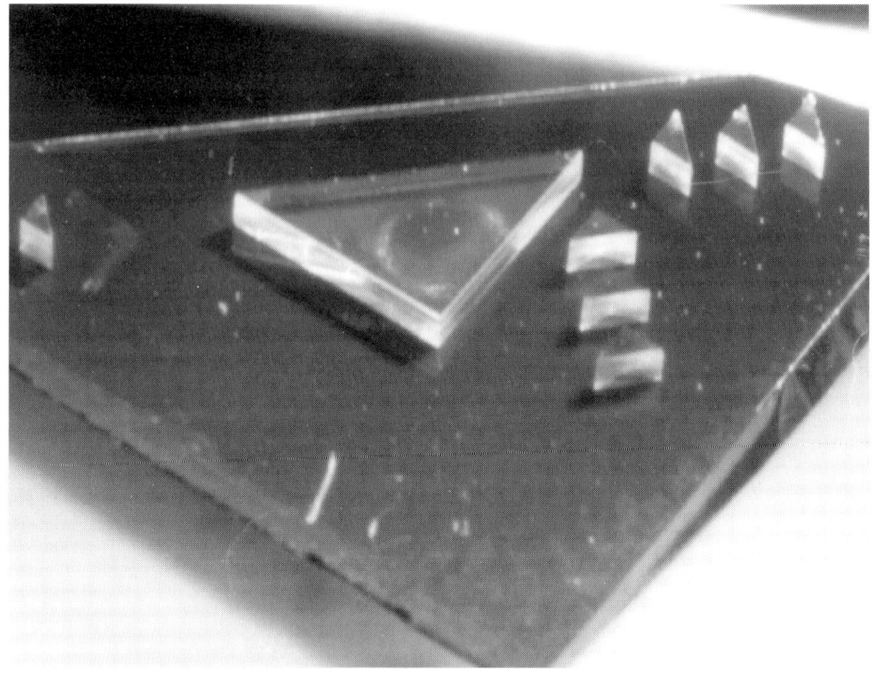

EXAMPLE: Electrostatic linear actuator

Switching of light by means of an electrostatic linear actuator has been reported in [Gött95b]. In a T-configuration light can be transferred to the opposite fiber or to the 90° fiber using a moveable mirror (fig. 9.11). By applying a voltage of approximately 48 V the mirror is moved out of the cross section. The switching time is 10 msec. To prevent resonant frequency the device is switched only in the low frequency domain (eg. 2 Hz). The maximum displacement is in the range of about 80 μm.

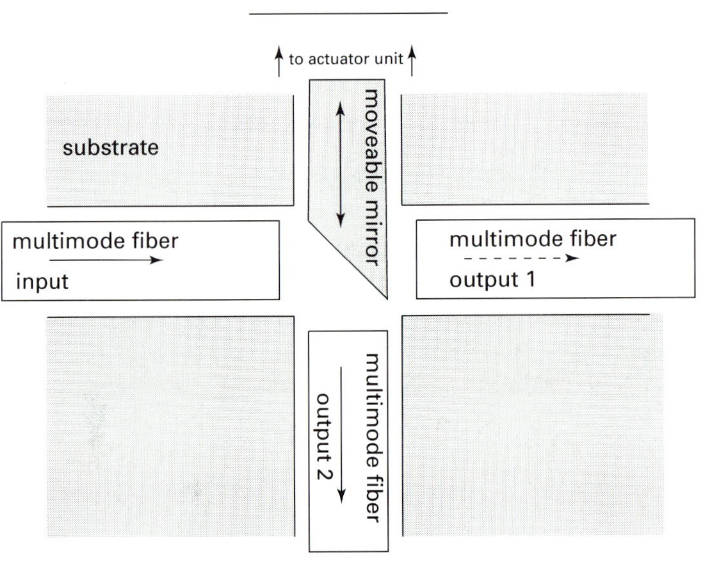

Fig. 9.11 – Multimode fiber coupling using a moveable mirror
(After [Gött95b])

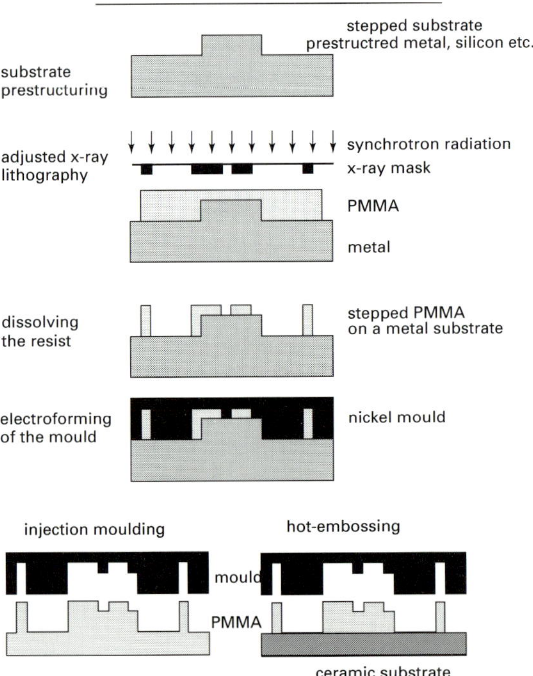

Fig. 9.12 – Fabrication process for micro-optical benches with several
level heights (After [Müll96])

9.3.2. Optical benches

For coupling of eg. singlemode fibers, a micro-optical LIGA bench can provide the accurate lateral alignment accuracy. But for a good coupling efficiency also microlenses are needed that fit into the LIGA bench. In order to adjust different structural heights eg. the optical axis of ball lenses with the optical axis of single mode fibers, LIGA can be performed on a silicon wafer [Gött95b, Müll95]. The fabrication process providing different structural height is shown in fig. 9.12. The following results have been optained for an optical test system consisting of two ball lenses and two single mode fibers.

ball lens diameter:	900 μm
fiber diameter:	125 μm
step height:	387.5 μm
vertical accuracy:	± 1 μm
lateral accuracy:	± 0.2 μm
insertion losses:	2 dB
principle losses:	1 dB (including Fresnel losses)

EXAMPLE: Bidirectional optical transceiver module

This principle of combining singlemode fibers and ball lenses has also been applied to an optical transceiver module shown in fig. 9.13 and 9.14 [Müll95, Müll96, Gött95b].

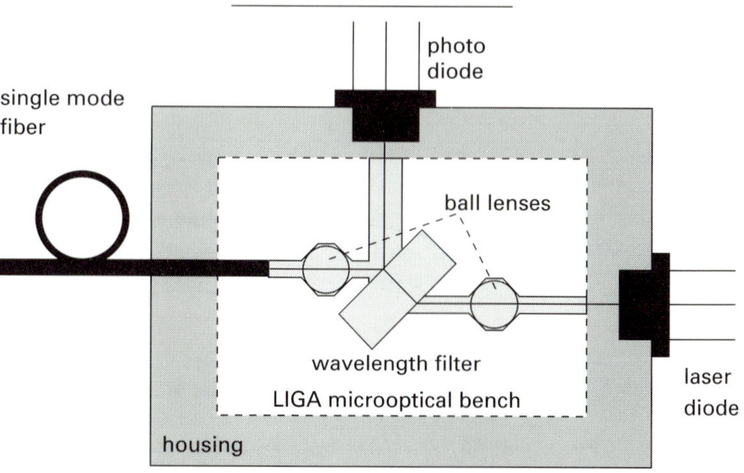

Fig. 9.13 – Schematic of a bidirectional transceiver module
(After [Müll95, Müll96])

Fig. 9.14 – Photograph of the bidirectional transceiver module
(courtesy A. Müller, Forschungszentrum Karlsruhe, [Müll95])

The bidirectional transceiver module is built up by a laser diode and a photo diode, a microopitcal bench with two ball lenses, a wavelength filter and a singlemode fiber. Light coming from the laser diode is collimated by the right ball lens, passes through the wavelength filter and is focused by the left ball lens into the fiber core. Light coming from the fiber is collimated by the left ball lens and is reflected by the filter onto the photo diode. The achieved characteristics are

insertion loss (laser diode - fiber):	-5.3 dB
insertion loss (fiber-photo diode):	-0.6 dB
cross talk:	-39 dB
return loss:	-16 dB

A similar concept with moving parts, fibers and collimating ball lenses for use as an optical 2x2 switching node has also been published [NN96].

EXAMPLE: Fiber-in-board interconnect

In [Vett95] a 'fiber in board' interconnection scheme has been reported. 16 optical graded index multimode fibers are embedded in an electro-optical board. The board is connected to an optical backpanel via two LIGA fiber connectors. To prevent a 90° outcoupling from the electro-optical board plane to the connectors, the fibers are embedded with a small angle as indicated in fig. 9.15.

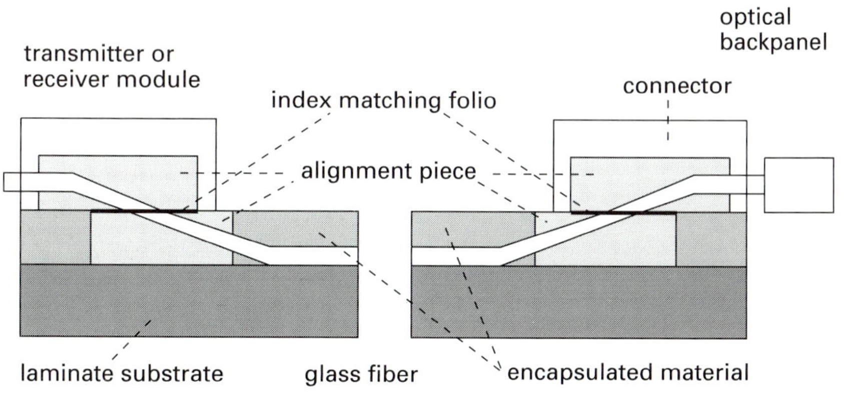

Fig. 9.15 – Schematic of fibers embedded in an electro-optical board with connectors (After [Vett95])

Precision alignment parts are integrated within the board to guide the optical fibers to the surface at an angle of 14°. The fiber pitch is 250 μm. After alignment and fixing the fibers with a polymeric resin, the upper surface is polished and an array of elliptical fiber interfaces is obtained. The connector consists of a LIGA precision plate made of nickel on top of a plastic submount. It allows the alignment of 16 fibers at a pitch of 250 μm according to the fibers in the substrate [McEn95]. Alignment rods and depth control parts provide accurate positioning.

The characteristics of the system are

attenuation surface optical contact: < 1 dB
cross talk suppression adjacent fibers: > 59 dB
modulation rate: 622 Mbits/s
bit error rate: < $1e^{-14}$

References for chapter 9:

[Beck82]   E.W. Becker, W. Ehrfeld, D. Münchmeyer: "Production of Separation-Nozzle Systems for Uranium Enrichment by a Combination of X-Ray Lithography and Galvanoplastics", Naturwissenschaften 69, 520-523, 1982.

[Beck86]   E.W. Becker, W.Ehrfeld, P. Hagmann, A. Maner, D. Münchmeyer: "Fabrication of microstructures with high aspect ratios and great structural heights by synchrotron radiation lithography, galvanoforming and plastic moulding (LIGA process)", Micro-electronic Engineering 4, 35-56, 1986.

[Bren93]   K.-H. Brenner, M. Kufner, S. Kufner, J. Moisel, A. Müller, S. Sinzinger, M. Testorf, J. Göttert, and J. Mohr: "Application of three-dimensional micro-optical components formed by lithography, electroforming, and plastic molding", Appl. Optics 32 (32), 6464-6469, 1993.

[Ehrf87]   W. Ehrfeld and E. W. Becker: "Das LIGA-Verfahren zur Herstellung von Mikrosturkturkörpern mit großem Aspektverhältnis und großer Strukturhöhe", KfK-Nachrichten. 19, 4/87, p. 167-179, 1987.

[Ehrf88]   W. Ehrfeld, P. Bley, J. Götz, J. Mohr, D. Münchmeyer, W. Schelb: "Progress in deep-etch synchrotron radiation lithography", J. Vac. Sci. Technol. B 6 (1), 178-182, 1988.

[Frit94]   A. Fritz: "Herstellung von Mikrolinsen aus PMMA durch Belichtung mit Synchrotronstrahlung und Styrol-Diffusion und ihre Charakterisierung", Dissertation, Uni Karlsruhe, 1994.

[Ghic81]   V. Ghica, W. Glashauser: "Verfahren zur spannungsfreien Entwicklung von bestrahlten PMMA-Schichten", Offenlegungsschrift DE 3039110.

[Gött92]   J. Göttert: "Grundlagen und Anwendungsmöglichkeiten der LIGA-Technik in der Mikroopitk", Thesis, Uni Karlsruhe, 1992.

[Gött95a]   J. Göttert, M. Fischer, and A. Müller: "High-aperture surface relief microlenses fabricated by X-ray lithography and melting", in Microlens Arrays, Vol. 5, 1995 EOS Topical Meetings Digest Series, p. 21 - 25, 1995.

[Gött95b]   J. Göttert, J. Mohr, A. Müller, and C. Müller: "Fabrication of micro-optical components and systems by the LIGA technique", Entropie, No. 192/193, p. 20 - 26, 1996.

[Heub88]   J. Heuberger: "X-ray lithography", J. Vac. Sci. Technol. B 6 (1), 107-121, 1988.

[Kufn93]   S. Kufner: "3D-Integration miniaturisierter refraktiver optischer Komponenten in PMMA", Thesis, Uni Erlangen-Nürnberg, Jan. 1993.

[McEn95] J. McEntee: "Microtechnology makes optics for industry", in Opto&Laser Europe, p. 36 - 38, April 1995.

[Mohr88] J. Mohr, W. Ehrfeld, D. Münchmeyer: "Requirements on resist layers in deep-etch synchrotron radiation lithography", J. Vac. Sci. Technol. B 6 (6), 2264-2267, 1988.

[Müll95] A. Müller, J. Hehmann, A. rogner, J. Göttert, and J. Mohr: "Hybrid optical transceiver module with a micro optical LIGA-bench", in proc of 21st Eur. Conf. on Opt. Comm. (ECOC'95- Brussels), p. 465 - 468, 1995.

[Müll96] A. Müller: "Aufbau hybrider mikrooptischer Funktionsmodule für die optische Nachrichtentechnik mit dem LIGA-Verfahren", Thesis Uni Karlsruhe, 1996

[Münc87] D. Münchmeyer, W. Ehrfeld, A. Maner, and W. Schelb: "Aufbau von Fertigungseinrichtungen für Masken für die Röntgentiefenlithografie", KfK-Nachrichten, 4/87, p. 180-191, 1987.

[NN96] "Optical switch facilitates flexible office networks", in Opto&Laser Europe, p. 38, April 1996.

[Schi94] H. Schift: "Herstellung und Untersuchung photonischer Mikrobauelemente in LIGA-Technik", Thesis, Uni Karlsruhe, 1994.

[Vett95] P. J. Vetter, J. Vandewege, G. De pestel, Q. Tan, F. Mignom, J. Van Koetsem, D: Morlion, A. Picard, A. Ambrosy, and G: Vendrome: "Optical interconnection using 'fiber in board' technology", in proc of 21st Eur. Conf. on Opt. Comm. (ECOC'95-Brussels), p. 445 - 448, 1995.

Deep Proton Lithography

Deep proton irradiation in poly (methyl methacrylate) (PMMA) is a fabrication method for monolithic integrated micro optics which allows structural depths in the order of several hundred microns. It offers high stability from the point of view that different optical functions can be fabricated in one block in order to form monolithic integrated optical systems, and in addition there are interesting mechanical autoalignment features. A variety of elementary refractive micro-optical components and monolithically integrated combinations can be fabricated: microlenses, microprisms, beam splitters, fiber connectors with self-aligned microlenses on top of each fiber.

10.1. Basic processes

The fabrication process consists of three basic elements: irradiation of a PMMA substrate followed by either a development of the irradiated regions or a swelling of the irradiated regions by organic vapour or both applied to different regions.

10.1.1. The irradiation process

The fabricaton process starts with the irradiation of linear high molecular weight PMMA-targets with protons (fig. 10.1). The target is masked by a metal structure which is either fully transparent or fully opaque to the proton beam (fig. 10.2). The high energy protons split the long molecular chains in the areas behind the transparent parts of the mask.

Micro-optics and Lithography

Fig. 10.1 – Irradiation of PMMA in order to create domains with reduced molecular weight

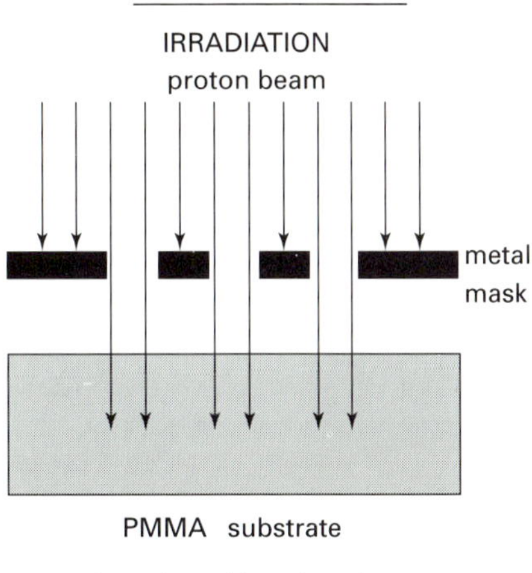

Fig. 10.2 – Photograph of a 300 μm thick Nickel mask made by the LIGA technique. Hole diameter is 1000 μm (test sample courtesy of Institut für Mikrostrukturtechnik, Forschungszentrum Karlsruhe)

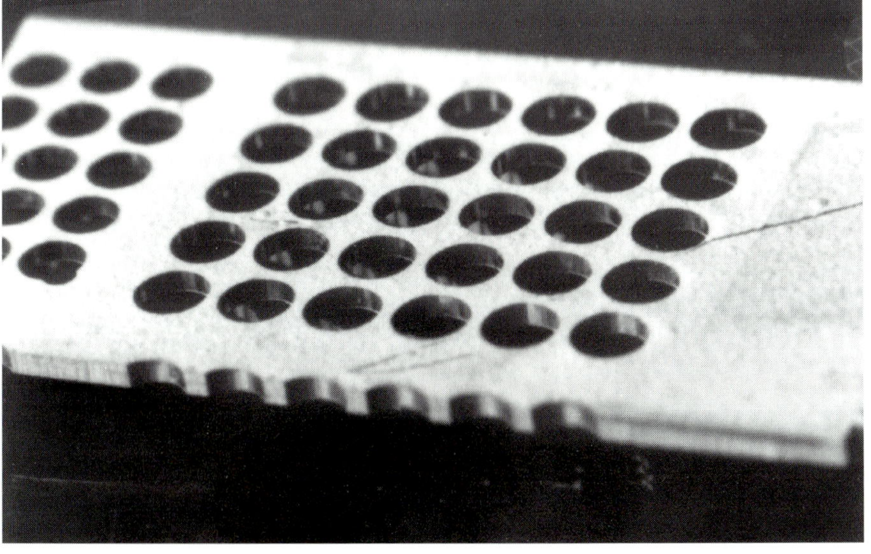

The intention of the irradiation process is to create well defined domains in a PMMA substrate with reduced molecular weight and thus with different chemical properties. The aspect ratio of the irradiated domains can be varied either by controlling the penetration depth or the size of the mask aperture. The penetration depth in PMMA is determined by the energy of the proton beam. The radiation damage resulting in a reduction of the average molecular weight can be controlled by the dose deposition.

Fig. 10.3 – Difference in energy deposition to PMMA between particle and electromagnetic radiation [Kufn96].

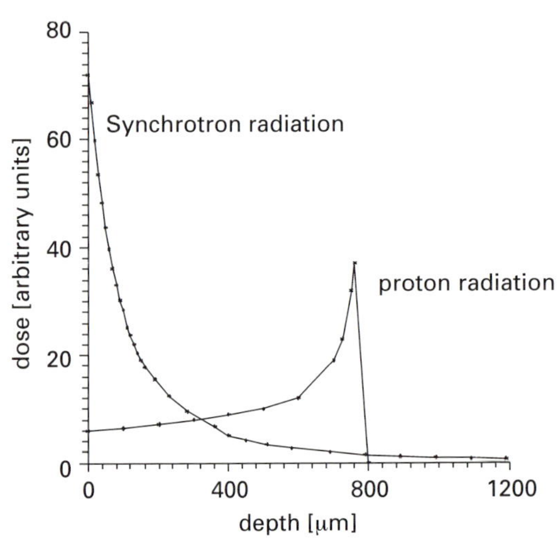

In the case of a proton beam the local distribution of molecular weight over the penetration range shows an accurate stopping depth of the incident protons corresponding to their kinetic energy. This is in contrast to the behaviour of electromagnetic radiation described in chapter 9, where the dose asymptotically decreases to zero, but never absolutely stops (see fig. 10.3.). The existence of a well defined stopping depth allows the creation of different structures on the two opposite sides of a PMMA plate without affecting each other. The energy transfer has a peak at the end of the penetration range when the speed of the incident particle reaches the order of magnitude of the speed of an electron [Jack75]. The longitudinal dose distribution therefore shows a wide zone with low variation and at the very end of the penetration range a peak with significantly higher dose deposition appears. Depending on the application it can be focussed on the one or the other effect.

Fig. 10.4 – Stopping depth of protons in PMMA, copper and gold as a function of their kinetic energy

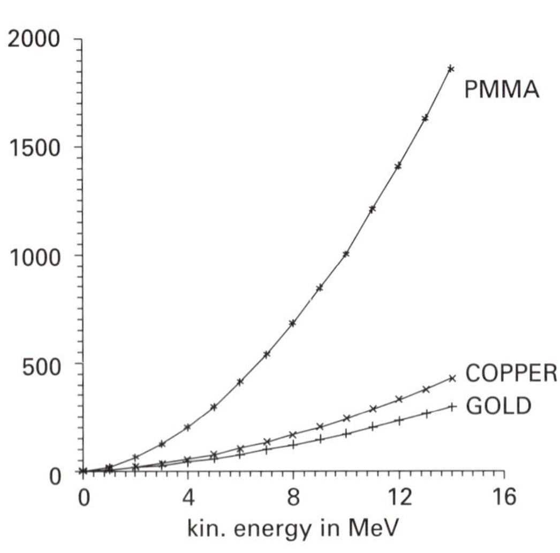

For deep etch lithography the achievable structural depth is very important. How far a certain particle can proceed before it is stopped, depends on its kinetic energy. The stopping depth of protons in PMMA as a function of the kinetic energy is shown in fig. 10.4. Penetration depths in the order of several hundred microns can be achieved for kinetic energies in the MeV range. Also depicted in this figure are the stopping ranges of protons in typical mask materials. Since a mask for proton irradiation must locally fully absorb the proton beam, the mask thickness must be more than the stopping depth in the mask material at the working energy used. Therefore the required mask thicknesses typically range in the order of 150 μm to 200 μm.

Besides the longitudinal effects of dose deposition there are also lateral effects which are of importance in the context of lithography. An effect of lateral straggling can be observed, since there are collisions of the incident particles with the particles of the penetrated material. A simulation of the proton irradiation in PMMA [Kufn96] shows the order of magnitude of the lateral straggling. The graph shows lines of equal dose deposition in a cross-section of a PMMA plate, irradiated homogenously with protons under normal incidence with an initial kinetic energy of 8 MeV. In fig. 10.5a the mask is a perfect slit, in fig. 10.5b a perfect edge. Once a PMMA sample is irradiated it can undergo a development or a swelling procedure. Each of these two processes can be applied independantly, but also combinations of the two are possible.

Fig. 10.5 – Lines of equal dose deposition in PMMA after proton irratiation a) over a slit, b) over an edge

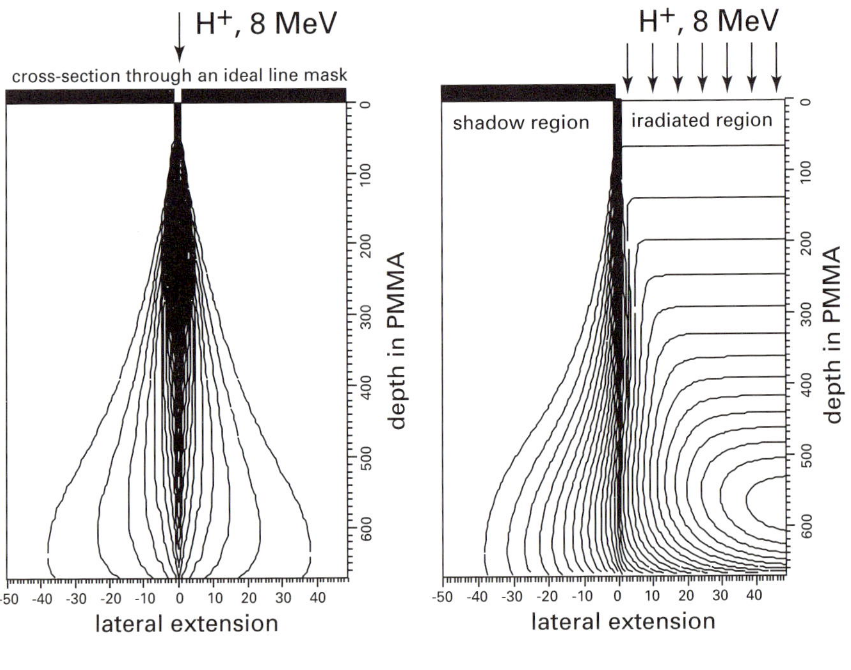

10. 1. 2. The development process

By irradiation of PMMA with protons areas with reduced molecular weight are created. In those areas the polymer matrix is more flexible and thus shows higher solubility than the non irradiated domains. These areas can be dissolved in a suitable developer (fig. 10.6).

In order to dissolve only the irradiated domains without significantly affecting the rest the developer has to be choosen carefully. Commercially available developers which are in use for lithography with positive and negative resists are not suitable for deep etch lithography. Their dissolution rate in non irradiated regions (dark rate) is in the order of 1 μm/s, which is not tolerable for development times in the range of 1 hour for the irradiated domains. In [Ghic81] a developer is described which was originally introduced for deep x-ray and electron beam lithography. It was possible to show that the same developer can be used for deep proton lithography if a purely linear PMMA with high molecular weight acts as resist material [Kufn90].

Fig. 10.6 – Dissolution of irradiated structures by development
(depth is several hundreds of micron)

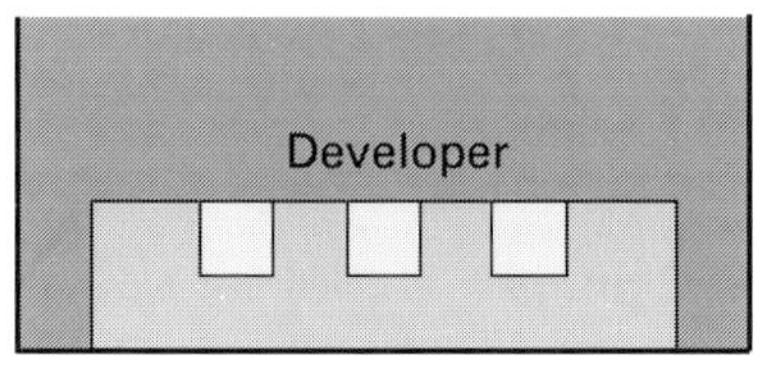

Fig. 10.7 – Monomer diffusion causes a volume expansion in the
irradiated domains.

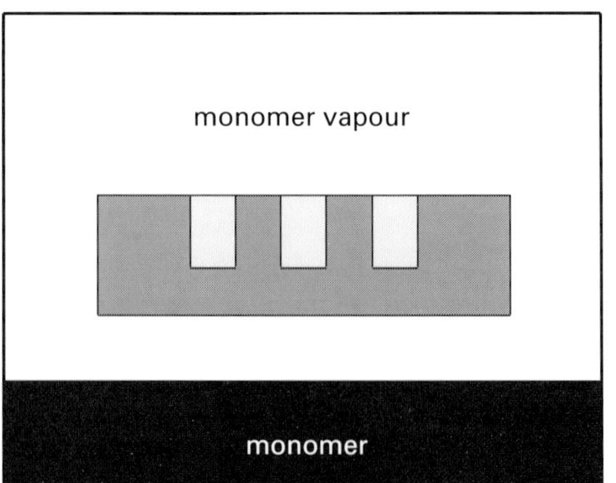

10. 1. 3. The swelling process

Alternative to the development a completely different procedure can be applied to the irradiated PMMA: the swelling of the irradiated domains. This technique makes use of the fact that the irradiation of PMMA changes also the swelling behavior of the PMMA matrix as a function of the molecular weight [Fran91].

After irradiation the target is exposed to styrene vapour typically at temperatures between 85° C and 95° C. The styrene molecules diffuse into the areas damaged by the proton beam. The shorter polymer chains in the irradiated areas integrate the monomer molecules by copolymerization. During the diffusion time almost no reaction between the long chains and the monomer molecules takes place. On the other hand, in the irradiated areas a considerable volume expansion can be observed which forms surface structures of lenslike shape (fig. 10. 7.).

The factor of volume expansion can be expressed by the parameter q, defined as the ratio of the initial volume V_0 to the expanded Volume V, the absolute volume growth ΔV is denoted as the difference between the initial volume V_0 and the final volume V (see fig. 10.8).

Fig. 10.8 – Definition of swelling parameters

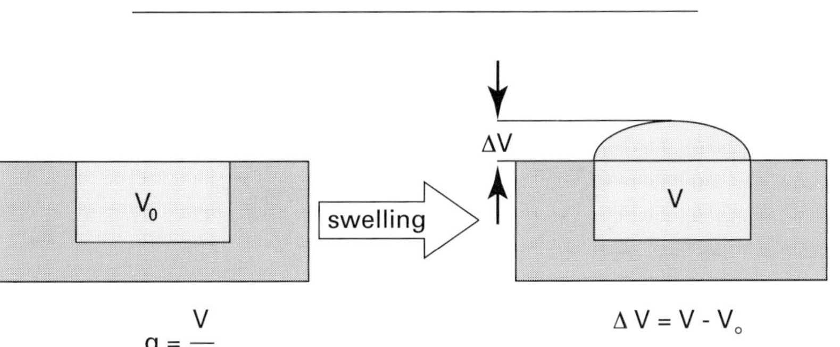

The swelling of polymers was already observed and analysed in 1953 from P. J. Flory [Flor53]. By insertion of other molecules into the matrix, for example by in-diffusion of a monomer, the matrix is expanded by the inserted material. On the other hand it is held together by the bondings within the matrix itself. These two forces in opposite direction balance to an equilibrium. Of main interest for the application of this effect to micro-optics or micromechanics is the relation between the molecular weight (which can be manipulated by irradiation) and the maximum achievable volume expansion. Quantitative reults are given by the Flory-Huggins theory [Flor53]. The swelling factor q is related to the maximum chain length M_c between two crosslink points in the polymer matrix by relationship (10.1). This means that the molecular chains can not arbitrarily be expanded, but there is an upper limit for the relative volume growth. This depends especially on the molecular structure of the polymer (represented by the parameter M_c) and therefore can be influenced by irradiation.

$$q^{\frac{5}{3}} \sim M_c \qquad (10.1)$$

Fig. 10.9 – The surface structures can be stabilized by photoinitiated polymerization

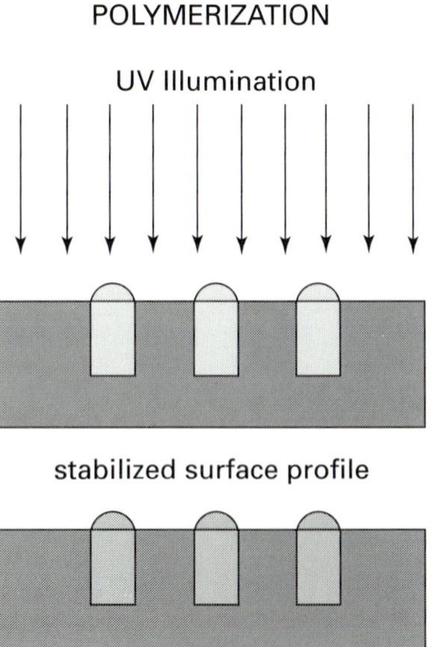

After diffusion these surface structures have to be stabilized in order to inhibit out-diffusion (fig. 10.9). This can be done by photoinitiated polymerization using UV light of 254 nm wavelength, which is highly absorbed by styrene, but not by polystyrene (see fig.10.10.)

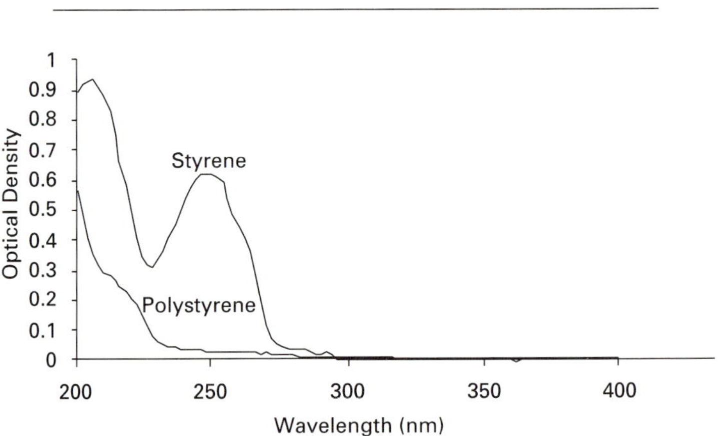

Fig. 10.10 – Absorption of styrene and polystyrene

10. 2. Fabrication of micro-optical elements

10. 2. 1. Fabrication of optical surfaces

The straightforward way of irradiation is a direct projection of the mask structure into the PMMA. But this means that every roughness or defect in the mask is also projected onto the developed structure. This results in grooves in the vertical surface of the PMMA structure. An example of these kind of grooves is given in fig. 10.11. It shows an interferometric measurement of the vertical surface of a developed structure. Any imperfections of the required thick metal masks concerning smoothness and verticality of the edges very critically affect the developed structures, which is critical if those are intended to be optical surfaces.

To overcome this disadvantage a variation of the method can be performed by moving the PMMA sample under the mask during irradiation. The mask however is installed in a fixed position relative to the beam. Moving the sample several times along the same path causes the roughness of the mask to be spatially blurred in the PMMA structure. Fig. 10.12 schematically depicts the different procedures.

The smoothening effect of this method can be seen most obvious for a mask of poor quality. So in the experiment shown in the following a mechanically drilled hole was used for a mask. The hole is situated in 250 µm thick copper plate with an aperture diameter of 125 µm. During irradiation the PMMA was moved several times forward and backward behind this mask using high precision motor drives.

Fig. 10.11 – Interferometric measurement of a surface irradiated under a fixed mask [Kufn93a]

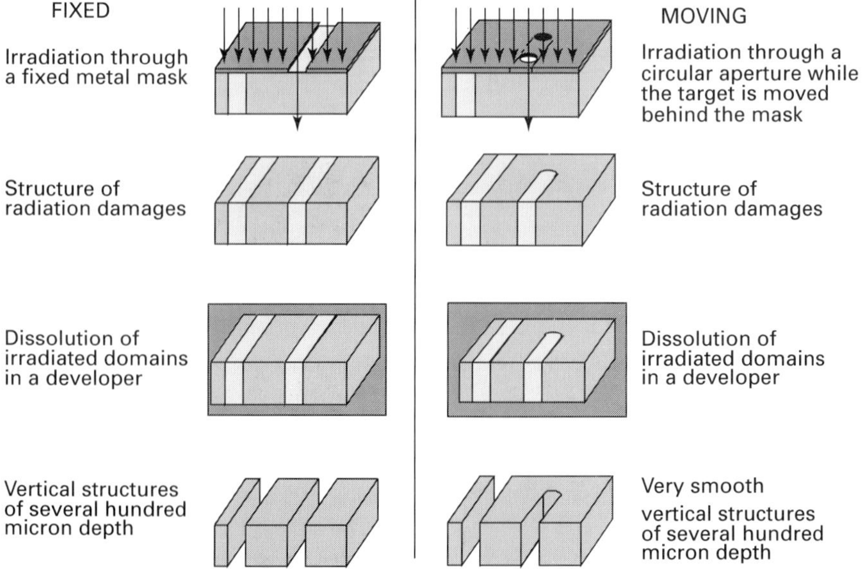

Fig. 10.12 – Irradiation of a slit structure: directly over a slit mask (left) or with the sample moving behind a pinhole (right) [Kufn96]

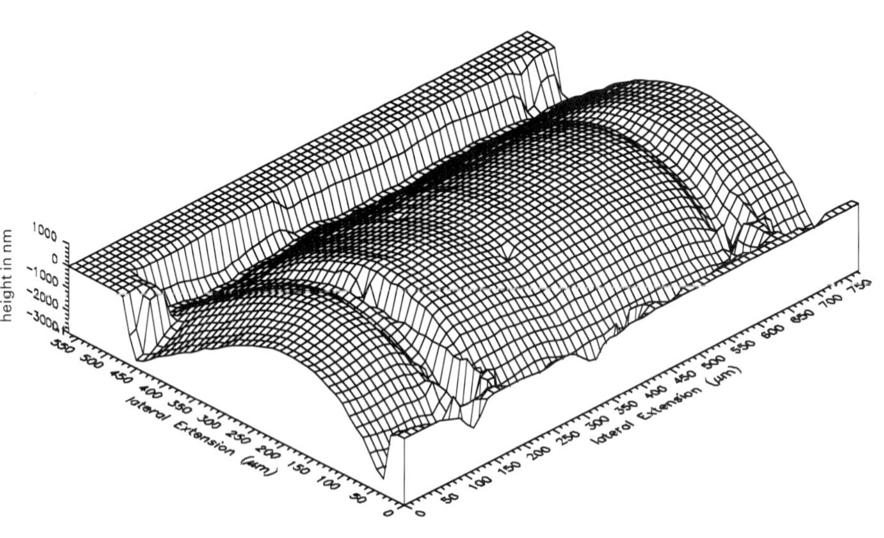

The corresponding interferogram of the surface is shown in fig. 10.13a. The curvature at the top and at the bottom of the structure is caused by the lateral straggling of the protons at the bottom and of the imperfect shape of the drilled mask at the top. In the central region which is zoomed in fig. 10.13b it can be seen that there the flatness is better than 1 μm. In the direction of the movement (horizontally) the roughness of the mask is blurred and a surface roughness of better than $\pm$ 20 nm can be achieved in the PMMA, as can be seen in the line scan in fig. 10.13c. This is especially important for deflecting surfaces such as prisms, beamsplitters and mirrors, but also for a combination of optics and mechanics in more complex optical systems.

Fig. 10.13 – Interferometric measurement of a surface irradiated moving over a distance of 500 μm behind a 125 μm circular aperture: a) full area, b) zoom of the center, c) horizontal line scan [Kufn96]

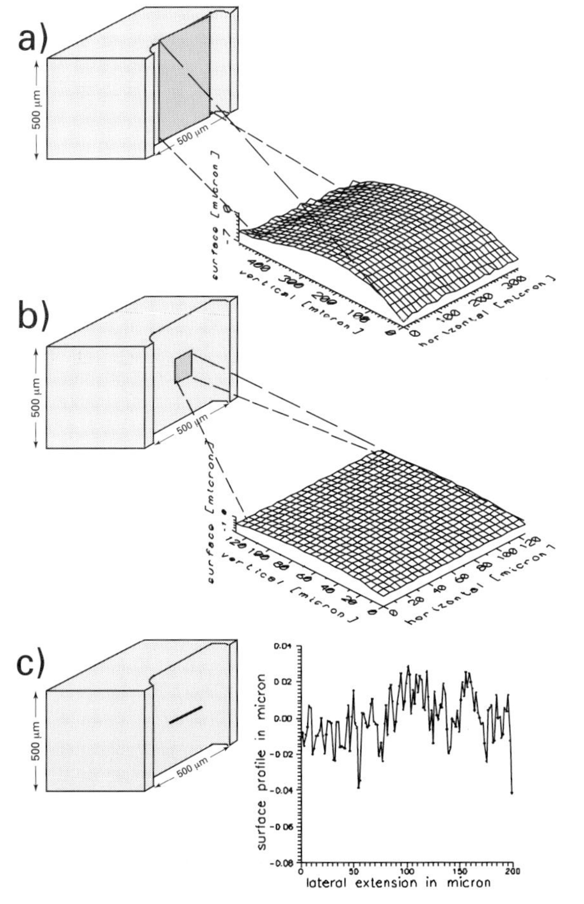

10. 2. 2. Fabrication of microlenses

The swelling effect of PMMA can be used to fabricate microlenses in the range of some hundreds of micrometer to about 2 mm. By irradiation through a circular aperture a cylindric volume with radiation damage is created. When this volume is swelled, the material can only extend via the top surface of the cylinder. Confined by surface tension this results lenslike shapes. The diameter corresponds to the mask aperture and the curvature to the swelling ratio. With sufficiently deep stopping range of the protons these lenses can have high numerical apertures, even half-spheres are possible. The photo in fig. 10.14 shows an example of a PMMA microlens with a diameter of 300 μm and a focal length of about 255 μm.

Fig. 10.14 – Photo of a PMMA lens with a diameter of 300 μm and a focal length of 255 μm [Kufn93b]

Arrays can be generated by a step-and-repeat process of irradiation and one subsequent swelling process. An example of an array of microlenses is shown in Fig.10.15. For reasons of surface tension the degree of freedom in the profile of the lenses is quite restricted. Variation from the spherical surfaces is only possible by changing the contour of the mask aperture thus obtaining profiles confined by e.g. ellipses or rectangles or hexagons. The photo in fig. 10.16 shows an array of hexagonal lenses irradiated in one step by a honey-comb mask. The hexagons have diameters of 80 μm with a separation of 8 μm.

Fig. 10.15 – Photo of an array of microlenses with diameters of 500μm and a spacing of 1mm [Kufn93b]

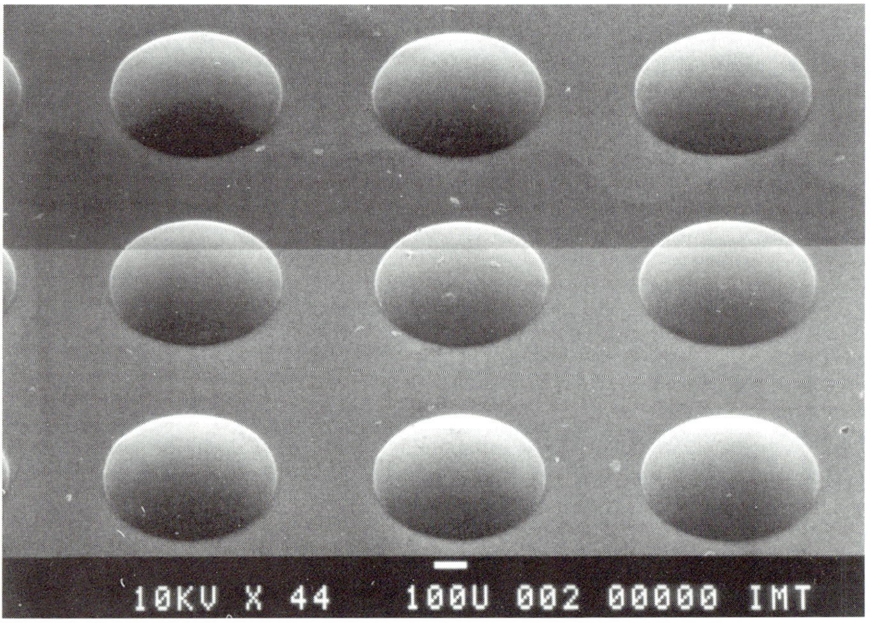

Fig. 10.16 – Photo of an array of hexagonal lenses [Kufn93b]

In the irradiation process the deposited dose allows control of the focal length and thus the F-number of the resulting lenses. The F-numbers achieved experimentally are in a range from 5 to 1 (half-spheres). The experimentally achieved lens diameters are from 50 μm to 1.8 mm (the availability of metal masks with sufficient thickness is the limiting factor for the achievable size). Imaging experiments with 500 μm diameter lenses show that line grids with a spatial frequency of 4 μm (256 line pairs/mm) can be resolved. The surface profile lenses are purely refractive and thus the wavelength dependence is significantly reduced compared to diffractive optics. The lenses can be copied easily by casting in the same way as all other surface lenses in the previous chapters. (There is also a slight change of the refractive index but the dominating effect results from the surface profile).

10.2.3. Fabrication of mechanical mounts

The procedure of cutting certain patterns from or into a PMMA sample can be used for mechanical purposes, too. Alignment between different building blocks can be improved by providing mechanical support structures as eg. with LIGA-structures (see chapter 9). This can be in the sense of cutting around an optical component, e.g. a lens array in order to obtain a precisely defined sample size such that it fits in a holder (fig. 10.17)

Fig. 10.17 – Fabrication of mechanical mounting support by cutting

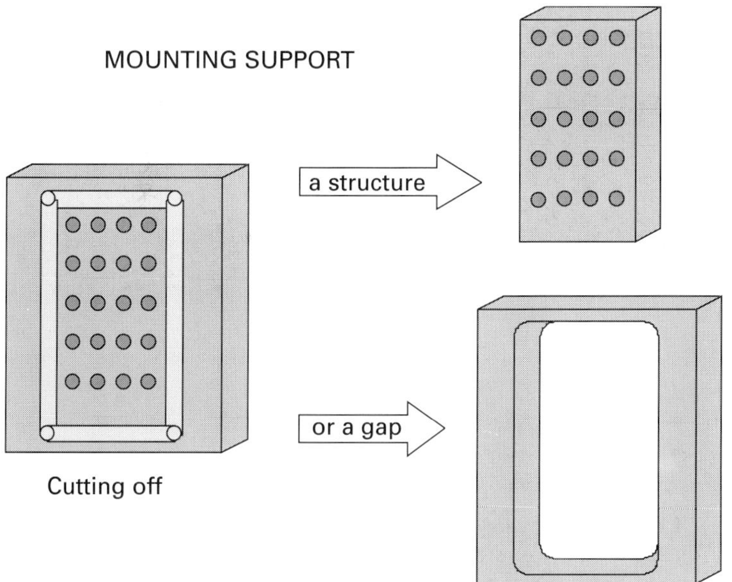

The complementary task, where the cutting techinque can be applied is providing the holder structure itself. Holder structures for three-dimensional systems can be fabricated in form of structured base plates where optical components can be mounted (fig. 10.18). Additionally mechanical holder structures for fibers can be fabricated by irradiation and subsequent development. This application will be discussed in more detail in combination of the development and swelling process.

In addition, the swelling effect can also be used for the mechanical purpose of shrinking a gap by swelling the holder itself. This will be treated in the chapter on the applications for optical fibers.

Fig. 10.18 – Base plate for mechanical mounting

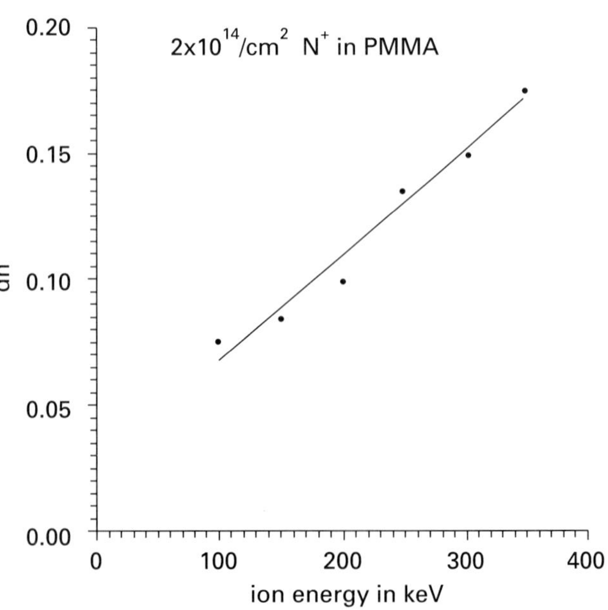

Fig. 10.19 – Nitrogen irradiation of PMMA: Change of optical refractive index, Δn, versus incident ion energy with a constant ion flux of $2 \times 10^{14}/cm^2$ (After [Kall91])

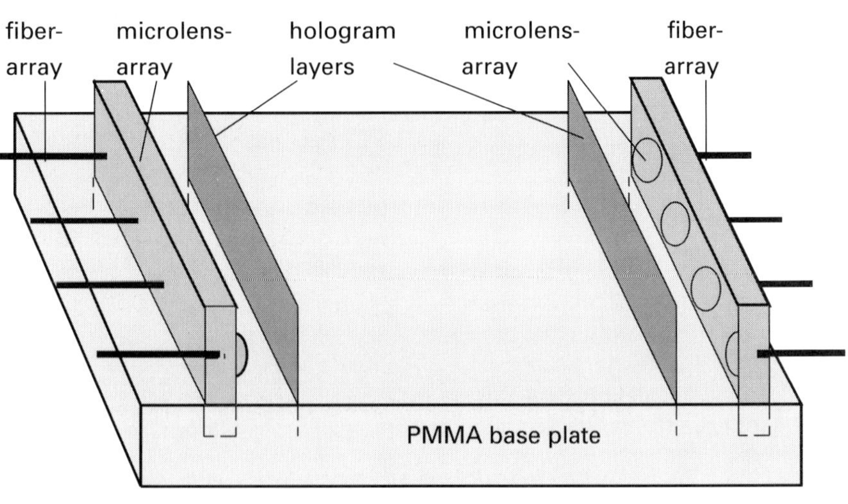

10. 2. 4. Waveguide fabrication

Stopping of protons, or in general, of ions is combined with high dose deposition in the stopping region. From the beginning of this chapter up to now only the molecular change was considered to be of importance. But there is also a significant change of the refractive index of the irradiated region which reaches a maximum arround the stopping depth. In [Kall91] an index change of $\Delta n > 0.1$ has been reported for low energy nitrogen ions in PMMA which is shown in fig. 10.19. A combination of index change and dose peek in the stopping region has been reported in [Fran93]. This method was used to fabricate buried waveguides in a depth of 35 μm in PMMA by means of alpha irradiation at 5.6 MeV.

10. 3. Applications to micro-optical sytems

10. 3. 1. Miniaturized deflection systems

Two small micro-optical systems which can be directly fabricated by the cutting method mentioned in chapter 10.2.1 are given as examples. Fig. 10.19 shows the principle and a photo of a Brewster telescope for six entries. In this setup the beam compression is only in horizontal direction. This can be used to reshape the eliptical output of a line of laser diodes into spherical beams [Lohm89, Kufn96].

Fig. 10.20 – Brewster array telescope for six entries [Kufn96]

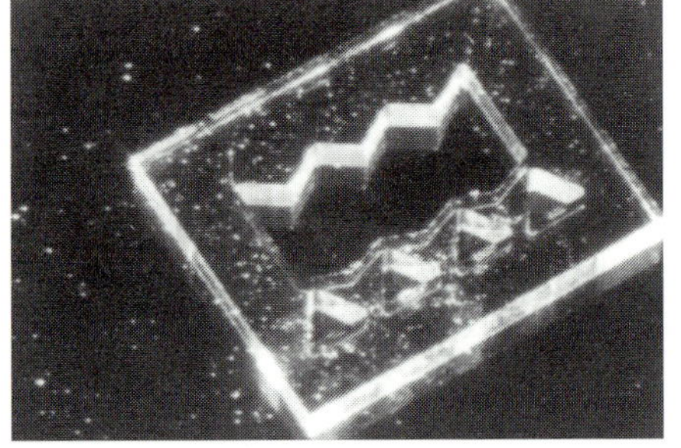

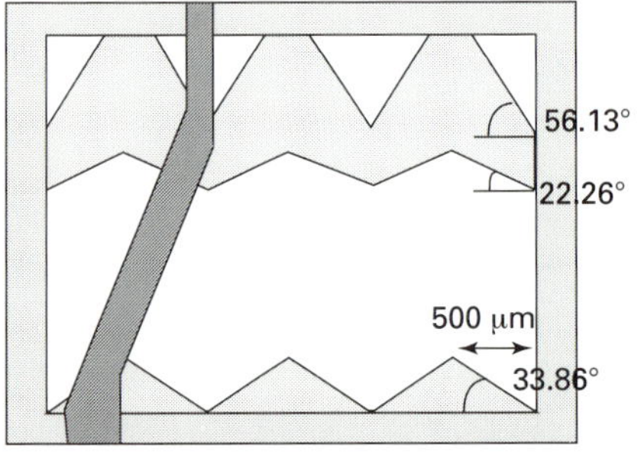

In fig. 10.21 a Perfect Shuffle permutation for 16 entries is shown that uses only refractive microprisms to provide a shuffle operation [Lai92, Kufn96].

Fig. 10.21 – Perfect Shuffle permutation for 16 entries [Kufn96]

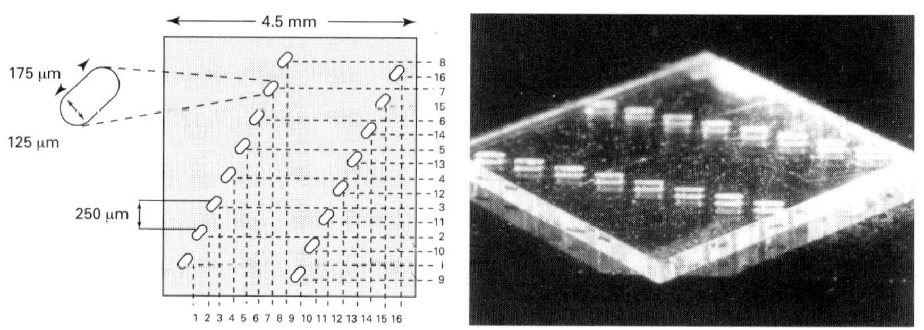

Fig. 10.22 – Fabrication of deflection prisms by tilting the substrate against the beam incidence

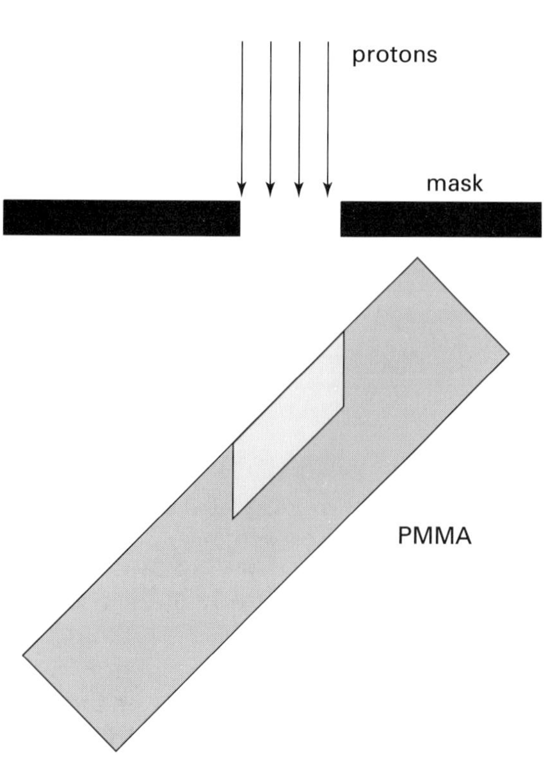

10. 3. 2. Miniaturized deflection- and imaging systems

The concept of dividing a system into a small number of simple and flexibly combinable modules is successfully used in electronics in the form of integrated circuits. With passive micro-optics the need for modules is not so obvious. But especially for the construction of complex free space optical systems with numerous lenses, prisms, beam splitters and also active devices a modular design can simplify the assembly significantly. In the following section a modular micro-optical system is constructed step by step. The used components consist of deep proton irradiation microelements integrated monolithically in one PMMA substrate [Kufn93a].

A basic component for a modular system is formed by an array of equidistant prisms with an alternating orientation of +/- 45° with respect to the substrate surface. The tilt angle of the prism is achieved by tilting the substrate relative to the beam incidence (fig. 10.22). Arrays of alternating ± 45° prisms can be fabricated by tilting the substrate to these angles. Fig. 10.23 shows a photo of a prism array fabricated in this way.

Fig. 10.23 – Array of alternating ± 45° prisms [Kufn93a]

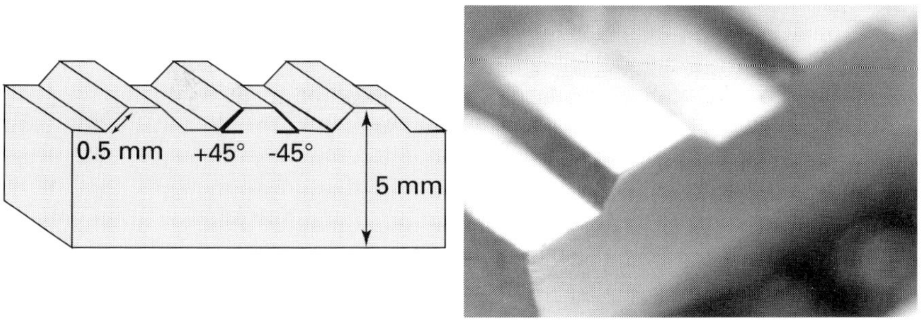

Also beamspitting can be achieved by metal or dielectric coating techniques applied to the microprism surfaces. Then beam coupling in or out is possible at selected locations. Fig. 10.24 shows a schematic view and a photo of a beam splitting experiment with a component which was coated with a 35 nm gold layer.
Additionally a microlens array can be integrated with a microprism array on opposite sides of one substrate. On the backside of the above prism array a microlens array can be generated in a second processing step.

Micro-optics and Lithography

The extended component with the lens array opposite to the prism array
is shown in fig. 10.25. In this component the ± 45° edges of the prisms
have a length of 500 μm and the lenses have a diameter of 500 μm.

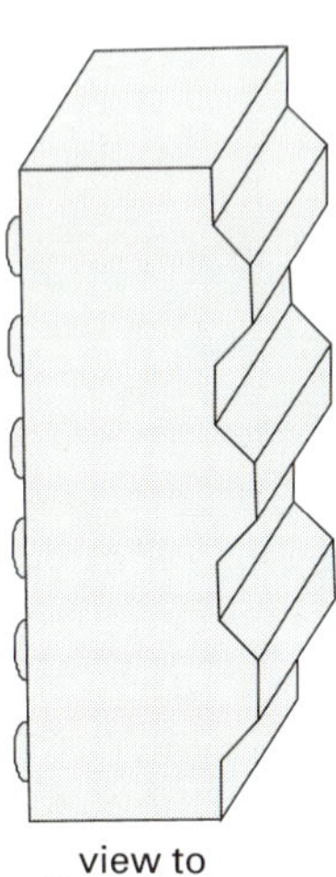
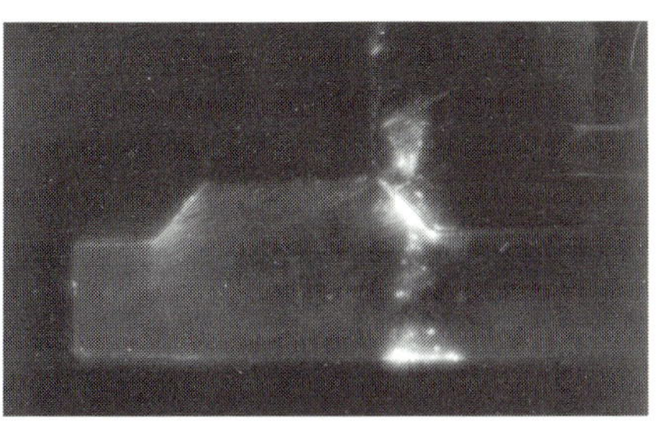

Fig. 10.24 – Beamsplitting experiment reallized by a 35 nm thick gold
layer between two PMMA blocks [Kufn93a]

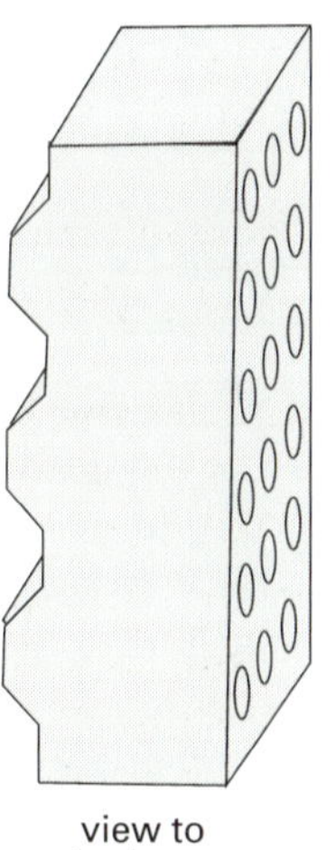
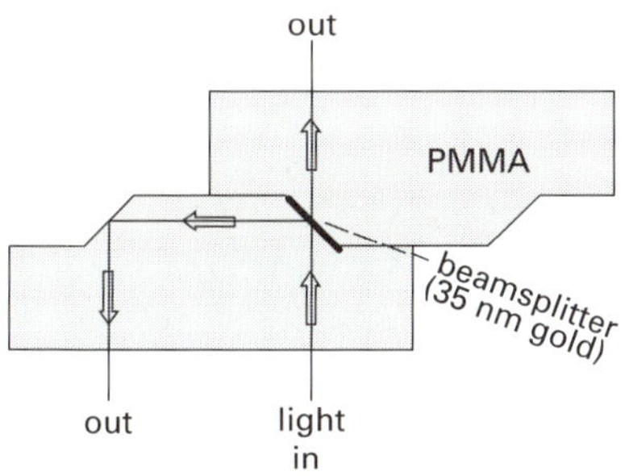

Fig. 10.25 – Monolithic integration of a microlens array opposite to a
microprism array [Kufn93a]

10. 3. 3. Three-dimensional fiber-lens connector array

For almost all fiber applications the combination of fibers and microlenses is of great relevance (see also chapter 6.2). A microlens can interact with a fiber by coupling light from a source into a fiber and by collimating light coming out of a fiber. In both cases the longitudinal and the lateral displacement between fiber and lens has to meet strict alignment tolerances especially if single mode fibers are required. Also of interest is the alignment between arrays of fibers and arrays of microlenses. Problems arise from the fact that conventional methods of fabricating a fiber holder and a microlens are quite distinct. For several years many different fabrication methods for microlenses have been published (see previous chapters). Due to the lithographic design methods it is easy to fabricate arrays of microlenses instead of only a single microlens. This is not the case for fiber holders. If several optical fibers have to be aligned, one line of fibers can be arranged in V-grooves (see chapter 4.4). Also integration by use of excimer laser has been published (see chapter 8.2). But even here there is still a need for a microlens array well adapted to the spacing and the tolerances of the fiber holder array.

Deep proton lithography is a tool for the fabrication of both microlens arrays and also for the fabrication of hole arrays which are suitable for fiber holding. A combination of these two kinds of arrays allows an assembly of fibers with lenses on top. Tolerances in the fiber diameter can be reduced by the possibility of self-centering of the fibers in the holder. For the fabrication of fiber holder arrays and microlens arrays a PMMA substrate is irradiated in a step-and-repeat process over a mask with a circular aperture corresponding to the size and pitch of the fibers. Important in this context is the distribution of molecular weight caused by the proton irradiation.

The simulation in fig. 10.26 shows lines of equal dose deposition in a PMMA sample of 500 μm thickness (which is in this range equivalent to the distribution of molecular weight). Dose deposition is determined by the energy dependance of the absorption coefficient and by scattering. An ideal edge is assumed to be irradiated homogeneously with N * 10^{13} particles per cm^2 with a proton energy of 7 MeV. The graph shows that at the bottom a significant dose deposition can be obtained in the geometrical shadow region of an edge which is caused by the lateral straggling of the protons. Furthermore it can be seen that the dose increases downwards which means that the maximum dose is deposited at the bottom of the structure. The choice of the dose parameter N is determined by the level where in the substrate the development processes can be stopped.

Fig. 10.26 – Dose distribution in PMMA under an ideal edge [Kufn95]

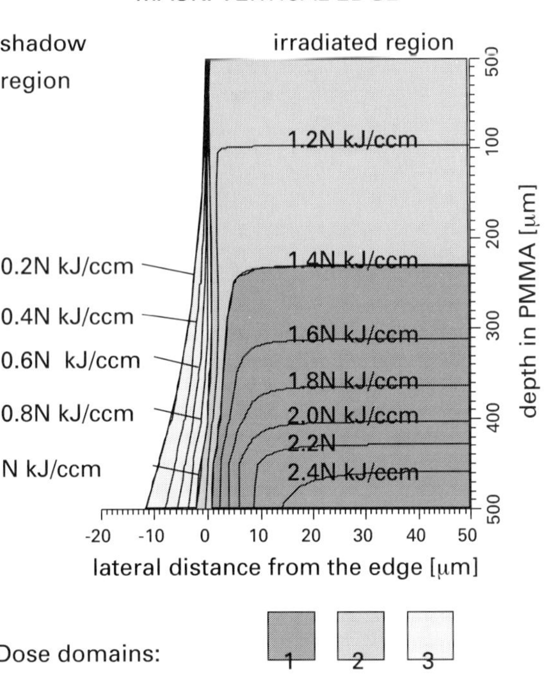

In fig. 10.26 three areas are labelled which are of interest with respect to applications for fiber optics [Kufn95]. For the fabrication of fiber holders the dose is choosen so that the developer can dissolve domains 1 and 2. This results in a fan-shaped hole through the substrate. A fiber can be plugged into this hole and afterwards the remaining domain 3 can be swollen in a styrene vapor athmosphere thus centering the fiber.

For the monolithic integration of fiber holders and lenses a lower dose is given to the substrate so that only domain 1 is developed. The resulting structure now is a blind hole into which a fiber can be plugged. A subsequent diffusion of styrene vapor will swell domain 2 and 3. The swelling of domain 2 forms a surface lens on top of the substrate well aligned to the fiber and as above domain 3 is swollen for centering and fixing the fiber in the blind hole. Fig. 10.27 shows the depth of domain 1 as a function of the dose in a PMMA substrate with a thickness of 500 μm.

Deep Proton Lithography

Fig. 10.28 is a side view of an array of plugged fibers with a monolithically integrated array of surface microlenses on top. The fiber diameter
is 125 µm. The substrate thickness is 500 µm.

Fig. 10.27 – Measured depth of development as a function of the dose
[Kufn96]

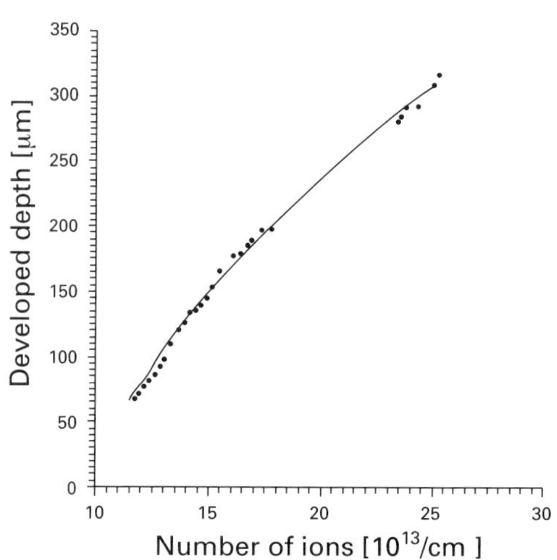

Fig. 10.28 – Monolithic integration of fiber holder and lens array
[Kufn96]

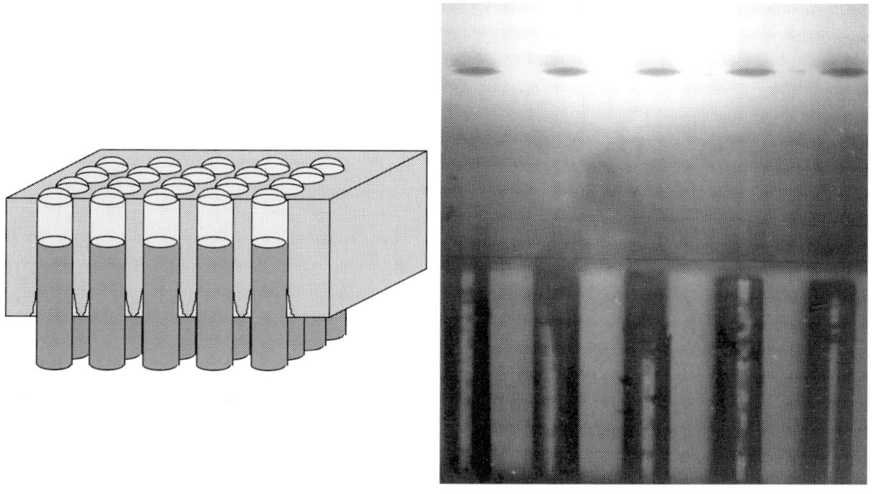

The method of deep proton irradiation allows the fabrication of fan shaped fiber holders which can compensate small tolerances in the fiber diameters by swelling the holder. Arrays of fiber holders and microlenses can be fabricated together in the same PMMA substrate. This kind of monolithic integration ensures that the microlenses are exactly on top of their corresponding fiber holders, therefore no further alignment is necessary.

References for chapter 10

[Bren90]   K.-H. Brenner, M. Frank, M. Kufner, S. Kufner: "H+-Lithography for 3-D integration of optical circuits", Appl. Optics 29 (26), 3723-3724, 1990.

[Flor53]   P. J. Flory: "Principles of polymer chemistry", Cornell University press, Ithaca, 1953.

[Fran91]   M. Frank, M. Kufner, S. Kufner, M. Testorf: "Microlenses in polymethyl methacrylate with high relative aperture", Appl. Optics, Vol 30, No. 19, 2666-2667, 1991.

[Fran93]   W. F. X. Frank, F. Linke, A. Schösser, T. K. Strempel, S. Brunner, D.M. Rück, U. Behringer, T. Tschudi, H. Franke, and T. Sterkenburgh: "Passive optical devices in polymers", in Proc. SPIE, Vol. 2042, paper 2042-30, 1993.

[Jack75]   J. D. Jackson: "Classical electrodynamics", 2nd edition, Wiley, New York, 1975.

[Kall91]   R. Kallweit and J. P. Biersack: "Ion beam induced changes of the refractive index of PMMA", Radiation Effects and Defects in Solids, Vol. 116, p. 29 - 36, 1991.

[Kufn93a] S. Kufner, M. Kufner, M. Frank, A. Müller, K.-H. Brenner: "3D integration of refractive micro-optical components by deep proton irradiation", Pure Appl. Opt. 2 (1993) 111-124.

[Kufn93b] M. Kufner, S. Kufner, M. Frank, J. Moisel, M. Testorf: "Microlenses in PMMA with high relative aperture: A parameter study", Pure Appl. Opt. 2 (1993) 9-19.

[Kufn95]   M. Kufner, S. Kufner, P. Chavel, M. Frank: "Monolithic integration of microlens arrays and fiber holder arrays in PMMA with fiber selfcentering", Optics Letters, Vol. 20, No. 3, 1995, 276-278.

[Lai92]   S. Lai and D. Hsu: "Perfect-shuffle implementation by using reflecting prisms", Applied Optics, Vol. 31, No. 29, p. 6191 - 6192, 1992.

[Lohm89]   A. W. Lohmann, S. Sinzinger, and W. Stork: "Array of Brewster telescope", Applied Optics, Vol. 28, p. 3835 - 3837, 1989.

Conclusions

11.1. Overviews of the technologies

The previous chapters introduced the main current technologies that already have or that will have influence on the future of micro-optics. Each technology has its very own strengths and weaknesses, and each covers a certain range of functionality. An overview of what the different technologies can provide is given in table 11.1.

11.2. Discussion of different approaches for one example project

This book started with an introduction into lithography as an important tool in the success of micro-electronics. It was also stated that micro-optics will find its way into computing systems by solving the interconnection problem. The aspect of optical interconnection in electronics could serve as a challenge for micro-optic technologies and every technology can now be asked how its solution to a specific problem would look like. Any choice of a project to discuss will certainly not cover all aspects in the field of miniaturized optics, but it should serve as a representative example to illustrate the numerous criteria to decide on. One example based on a project which is actually under investigation will be discussed in the following. This problem is known in literature since many years (see eg. [Jewe90]). But now a minaturized and integrated optical solution is expected.

Table 11.1

	diffractive (planar) optics	guided wave optics	gradient index optics	photosensitive glass	melting of photoresist	excimer laser	LIGA	protons
collimation ($\varnothing$, F/#)	$\varnothing$ > 100µm F/# > 2.5	-	10µm<$\varnothing$< 1000µm 2<F/#<25	75µm<$\varnothing$< 1000µm 1.5<F/#<25	10µm<$\varnothing$< 2000µm 1<F/#<15	10µm<$\varnothing$ 1000µm 5<F/#<12	50µm<$\varnothing$<< 1000µm 1<F/#<15	50µm<$\varnothing$< 2000µm 1<F/#<5
deflection	grid	bended guide	set of 2 lens arrays	set of 2 lens arrays	set of 2 lens arrays	refr. prism	refr. prism	refr. prism
beam splitting	grid	Y-branch	-	-	-	-	coated prisms	coated prisms
mechanical support	alignment marks	alignment marks	alignment marks	pedestals & marks	alignment marks	pedestal, microbench, fiber holder	pedestal, microbench, fiber holder	pedestal, microbench, fiber holder
replication (in plastics)	yes	partially	-	yes	yes	yes	yes	yes
monolithic integration	collimation, deflection, beamsplitt., array gen.	dir. coupler, Y-branch, active switching	collimation	collimation	collimation	collimation, deflection	cyl. lenses, deflection, beamsplitt., microbench	collimation, deflection, beamsplitt., microbench
dimensions (lat, struct. depth)	lat: < 1 µm depth: < 5 µm	lat: < 1 µm depth: < 5 µm	lat: > 10 µm depth: < 300 µm	lat: > 50 µm depth: < 2000 µm	lat: > 10 µm depth: < 100µm	lat: > 10 µm depth: no limit	lat: > 5 µm depth: < 2000 µm	lat: > 20 µm depth: < 1500 µm
material	glass, Si, plastics	glass, Si, LiNbO$_3$, plastics	glass, plastics	glass	photoresist, etching in glass	plastics	plastics	plastics
problems, potential	wavelength sensivity	1-dim. structures	flat surface	shrink of material	easy processing	excimer laser necessary	synchr. source necessary	proton accelerator necessary

Conclusions

Fig. 11.1 – Schematic of the task to be solved: An array of emitters must be interconnected to an array of detectors

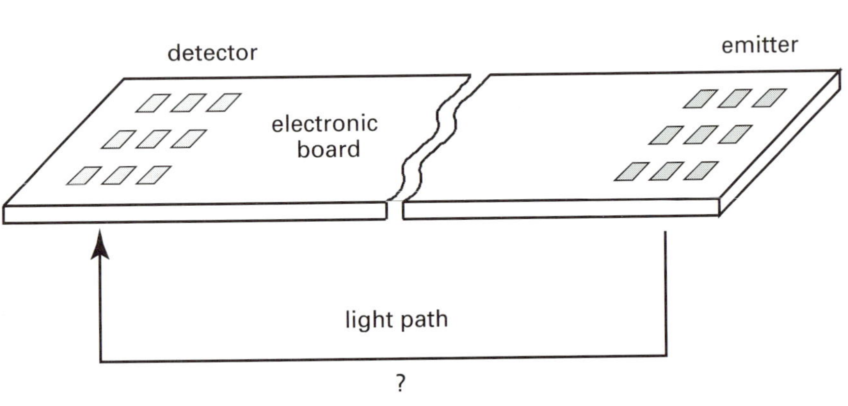

An electronic circuit board is assumed to carry a two-dimensional emitter array (eg. VCSELs, LEDs, PnpN-photothyristors etc.) and separated from it — on the same board (intrachip) or on a neighbouring board (interchip) — there is a two-dimensional array of photodetectors. For reducing the degrees of alignment between emitter and detector plane both arrays are situated in the same plane. A system has to be designed which directs the light from the emitter array to the detector array. A sketch of the configuration is shown in fig. 11.1. Based on the different technologies different approaches for both intra- and interchip interconnection can be envisaged.

11. 2. 1. Diffractive approach

A practical solution to the problem using diffractive optics can only be achieved with the planar optics approach. There all elements can be integrated in a single substrate with accurate alignment, and no further alignment stages are necessary. It should be mentioned that for planar optics it makes no difference in principle if emitters and detectors are placed on the same side of one substrate or in a stacked system, where emitters and detectors are placed parallel to each other. A solution to the problem of fig. 11.1 has already been published [Jahn93]. The schematic from that publication is shown in fig. 11.2.

Fig. 11.2 – Hybrid imaging system consisting of microlenses and macrolenses (After [Jahn93])

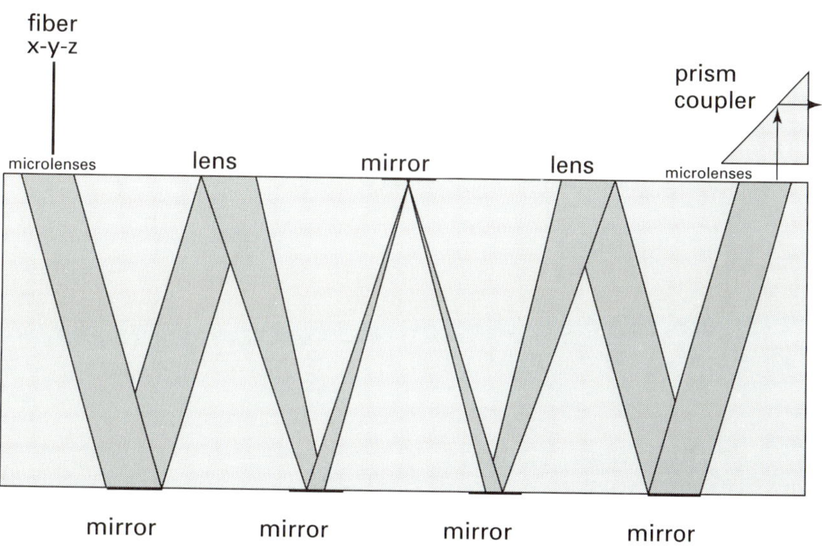

The system is a hybrid imaging system which uses microlenses for light collimation, and imaging is provided by two macrolenses. In this publication a fiber has been moved over the input microlens array and a video camera observed the output after the second microlens array. The planar block without the supplementary fiber and prism coupler can act as a setup for an integrated intrachip interconnection. The number of elements, the substrate thickness and the number of bounces between the substrate surfaces determine the overall length of the setup and the minimum separation between emitter and detector array. Note that the microlenses are designed with individual focal length.

An interchip interconnection with the same technique could look like the setup in fig. 11.3. To make the connection flexible two substrates are necessary which could be connectorized e.g. by a fiber ribbon. For the connector alignment marks and pedestals are needed. The light has to leave the substrate opposite to the entrance in order not to touch the electronic circuit with the fiber bundle.

Conclusions

Fig. 11.3 – Connectorized version for interchip interconnection using planar optics

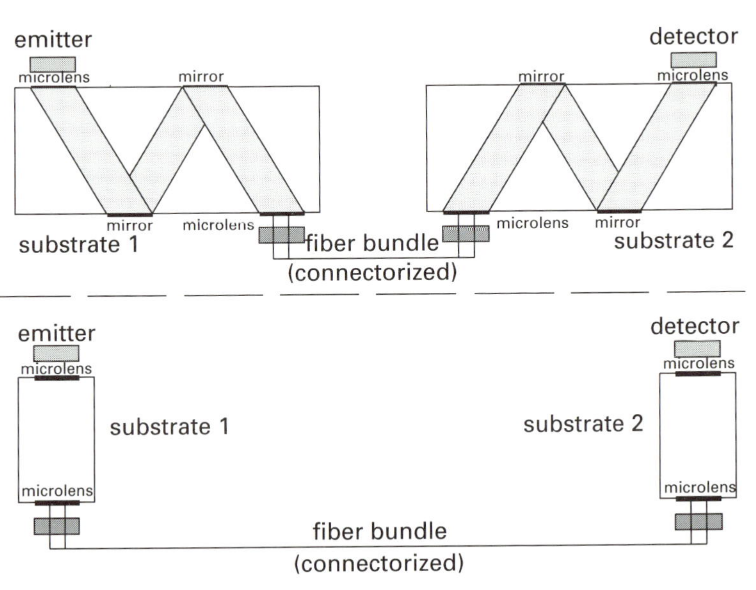

Since the fiber bundle and the connector cannot be fabricated with the technique of planar optics the interchip connection must use a hybrid approach.

11.2.2. Guided wave approaches

In chapter 9.3.2 (fig. 9.15) one appoach to this problem has already been mentioned. But it is a hybrid system which combines different techniques, such as waveguides together with connectors fabricated with deep etch lithography. The main problem in this case is the restriction of all guiding channels situated in one plane. A possibility to adapt to this geometry is a transformation of a two-dimensional array into a one-dimensional array as it is outlined in fig. 11.4. The connector can be in contact to tilted waveguides or to fibers embedded in a substrate. Dependent on the connector layout this represents a solution for both intrachip and interchip interconnect.
An alternative is shown in fig. 11.5. using the principle explained in chapter 4.2 (fig. 4.4). Embossed waveguides can also be fabricated in a layered structure with several layers on top of each other. The whole substrate can be polished at the end faces to achieve total internal reflection in the substrate.

Fig. 11.4 – Transition from a two-dimensional
to a one-dimensional array

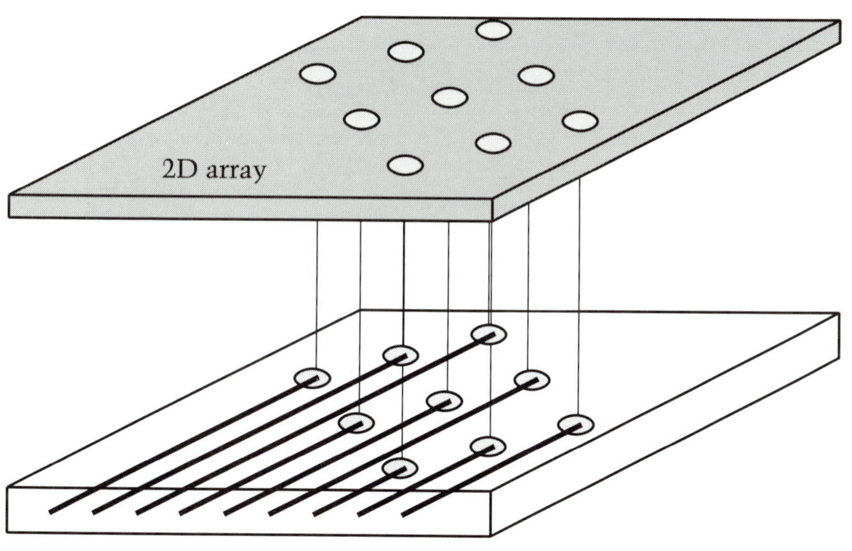

Fig. 11.5 – Monolithic guided wave solution using embedded
multimode waveguides and polished prisms

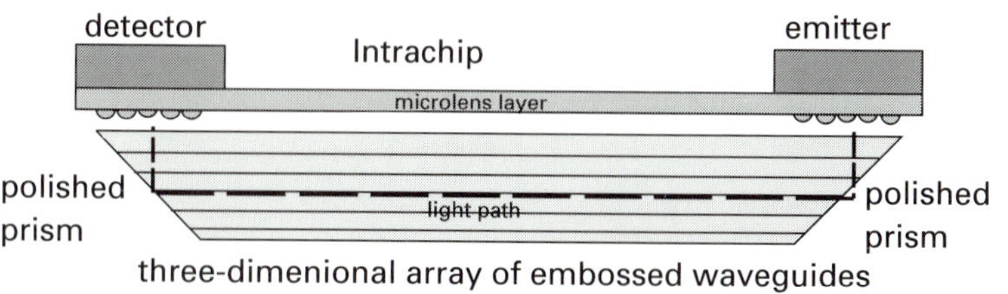

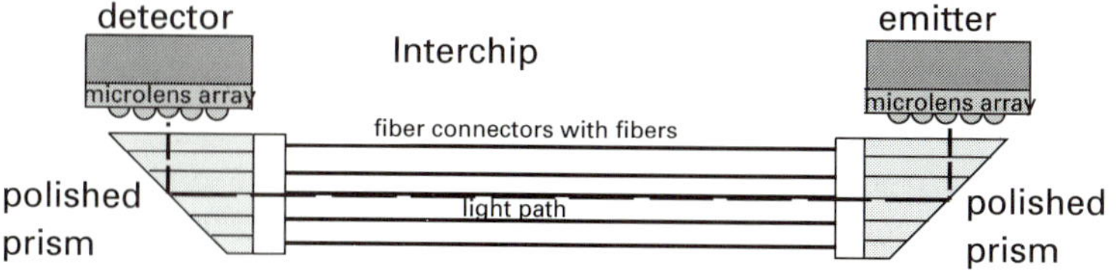

Conclusions

This yields a very compact monolithic block with guiding layers and deflection elements. Additional refractive or diffractive microlenses can be placed on top of the active elements to provide high light efficiency. Different to the diffractive approach in fig. 11.2 all microlenses have the same focal length. The method suffers from high tolerances because the different subsequent fabrication steps eg. embossing of the wave-guides and polishing of the prisms, have to be aligned relatively to each other. For multimode guides with high numerical apertures the setup can allow tolerances in the range of ± 5 μm which can be achieved with those techniques.

11.2.3. Refractive approaches

All fabrication methods described in the chapters 5, 6 and 7 are suited to provide light collimation and focusing elements which are indispensible to achieve high efficiency in the setups. Besides their optical perfor-mance there are differences in the mechanical and thermal stability and in the potential to integrate more functions than simple collimation.

Gradient index components

Glass microlenses in general have a better thermal coefficient than plas-tic materials which makes them preferable for those applications where temperatures higher than 80°C are expected. An advantage of the flat microlenses of chapter 6 is that neighbouring layers may touch any sur-face of the lens substrate which facilitates alignment. However the prims required for the connection setup have to be fabricated separately from the lenses and aligned to them. The idea of fig. 6.11 can be used to fabricate an integrated fiber-lens connector for interchip interconnec-tion.

Photosensitive glass

With the method of photosensitive glass tilted surfaces can be fabricat-ed, but the optical quality is not sufficient. Polishing techniques can help to overcome this problem or a glass substrate can be cut with a saw under an angle. In any case prisms and lenses have to be aligned. Thus the hybrid approach of fig. 11.5 has to be envisaged also for these tech-niques. For intrachip communication the waveguides are not strictly necessary. But for interchip communication the fiber approach is indis-pensable.

Eximer laser structures

Fig. 11.5 represents also a solution for excimer lasers. It is possible to fabricate ablated lenses or swelling lenses on one side of the substrate and to ablate a tilted prism on the other side. Depending on the optical quality of the deflection area it will be necessary to polish the prism after ablation. This technique can be used for a monolithic intrachip connection. For interchip communication fiber holders and additional holder structures can also be fabricated in the monoblock.

LIGA structures

LIGA can fabricate tilted surfaces with high optical quality and microlenses over a wide range of diameters and focal lengths. It can also provide hight accurate alignment structures, fiber connectors and optical microbenches. Using the layout of fig. 11.5, LIGA has to fabricate all elements separately and afterwards they are aligned precisely in a microbench. Using free space optics the intrachip solution can work without waveguide structures as in the case of the previous two paragraphs.

Deep proton lithography

Deep proton lithography has the possibility to provide monolithic fabrication throughout the setup. Furthermore in this approach tilted surfaces with high optical quality can be fabricated and the precision of the fiber holder is also suited for a singlemode fiber array if necessary. The monolithic version of fig. 11.5 is shown in fig. 11.6 and fig. 11.7. A monolithic integration reduces alignment tolerances significantly compared to the other approaches.

Conclusions

Fig. 11.6 – Intrachip communication with deep proton irradiation

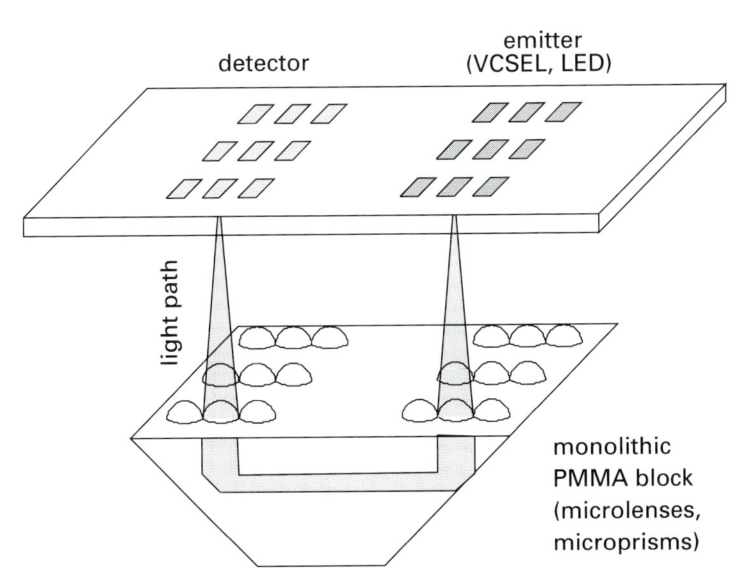

Fig. 11.7 – Interchip communication with deep proton irradiation

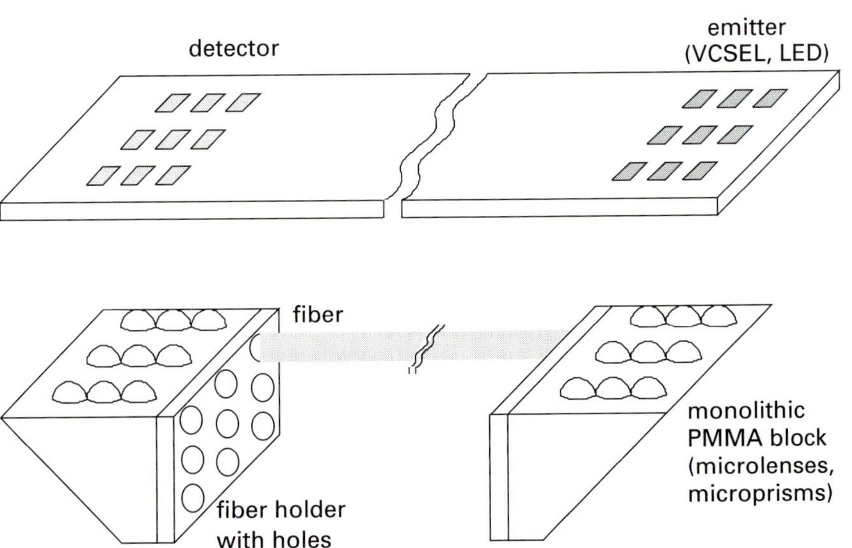

11. 3. Summary

In the above paragraphs different solutions to the problem of fig. 11.1 have been presented. Every technology can provide a solution to intra-chip and interchip interconnects. In most cases it will be hybrid. From that point of view none of them is best suited. A choice will mainly be made on criteria such as availability of certain techniques, alignment tolerances, temperature tolerances, performance of the required functions, and so on.

However micro-optical systems are much more than simply a collection of single components. The way in which these components fit together is especially important. In addition to the potential of each technology the compatibility of interfaces and the alignment effort must be taken into account when designing a system. Therefore monolithic or building block approaches are preferable to systems which are built from many individual parts. The requirements of each project must be examined. It may be found that these will disqualify some technologies that are otherwise very attractive.

References to chapter 11:

[Jahn93]   J. Jahns and B. Acklin: "Integrated planar optical imaging system with high interconnection density", Optics Letters, Vol. 18, No. 19, p. 1594 - 1596, 1993.
[Jewe90]   J. L. Jewell, Y. H. Lee, A. Scherer, S. L. McCall, N. A, Olsson, J. P. Harbison, and L. T. Florez: "Surface-emitting microlasers for photonic switching and interchip connections", Optical Engineering, Vol. 29, No. 3, p. 210 - 214, 1990.

Appendix:
Testing of Micro-optical Elements

Several techniques for the characterization of micro-optical elements exist. Depending on the optical functionality different aspects of the elements are important. For diffractive optical elements the etch depth and the lateral dimensions have to be known. Concerning refractive microprisms their optical flatness and roughness is important. In the case of microlenses their shape as well as their imaging characteristics are of interest. In this appendix several measuring tools will be presented which provide information about different aspects in micro-optics.

A.1. Measurement of the surface

If the optical element under test has a homogenous and well known refractive index, a measurement of the surface profile is sufficient to get the full information about the element. In case of rotational symmetry of an element even a line scan through the center characterizes the element. Some commonly used methods for suface measurement are listed in the following:

A.1.1. Line Scans

Microscope

A very simple method to get height information for a surface microlens can be observing the microlens from the side as shown in fig. 10.14. The contour of the lens represents a line scan over the top. This method does not contact the object and is very easy to carry out if it is assumed that a microscope is more often available in laboratories than e.g. a stylus profiler. Of course this only works for elements which are accessible for a side-view (which is not the case for arrays) and which contain no concave structures.

Stylus profiler

In a profiler a small-tipped probe is moved in a line across a surface like indicated in fig. A. 1. The stylus is made of a hard material (e.g. diamond) and has a tip radius of curvature between 0.1 and 25 µm [Crea92]. The load of the tip can be changed in order not to damage the test surface. But the stylus always has to be in contact with the test surface and soft materials like plastics can suffer from scratches. The dimensions of the stylus can keep it from going to the bottom of deep structures. Very often the maximum height level is smaller than 100 µm. The whole system must be vibration isolated. A stylus profiler can give very accurate results in lateral resolution (submicron range) and also in depth resolution (nanometer range). For the determination of a lens sag as shown in fig. A. 2. many measurements may be required to find the symmetry center, because the stylus can only scan a single line at a time, and this may not be necessarily the center of the lens. It is also a very effective method for diffractive optics because the depth profile which is important especially for phase elements can be measured with very high resolution.

Fig. A.1 – Schematic of a stylus profiler (After [Crea92])

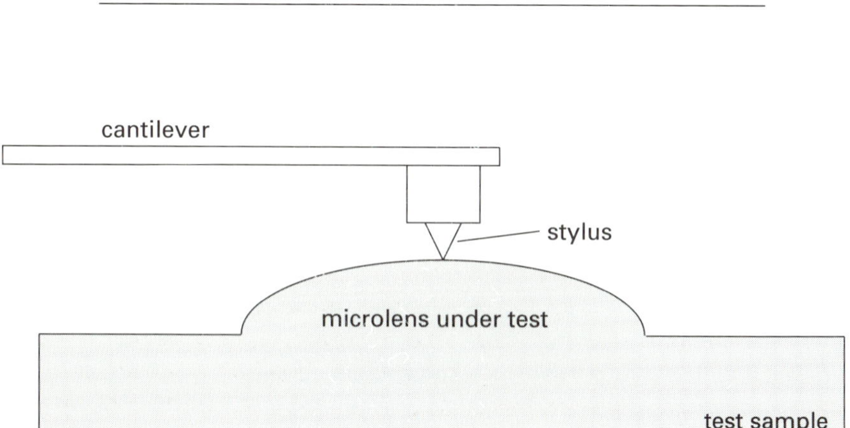

Fig. A.2 – Surface scan of a microlens by a stylus profiler [Laza96]

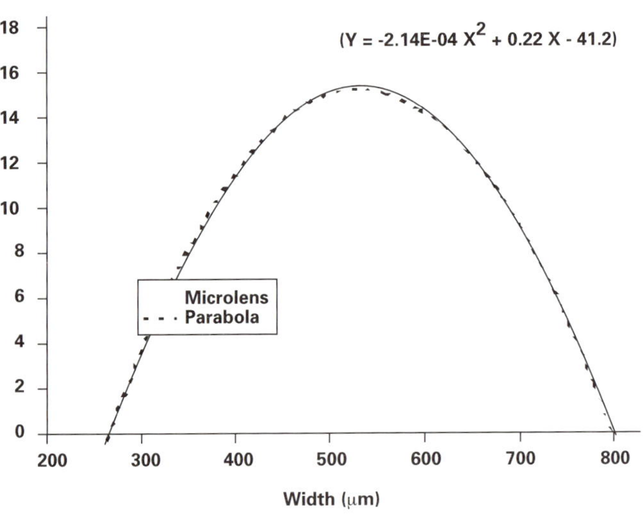

A.1.2. Optical focus sensors

An optical focus sensor scans the surface under test in two dimensions. In every position it determines the optimum focal position. Two different principles can be applied: differential electronic measurement or interferometric measurement.

Differential optical focus sensor

A differential photodiode system is similar to those in CD players where the height difference to a reference is recorded for every position. Thus, step by step the surface profile is calculated. A schematic of a focus sensor can be seen in fig. A. 3. The lateral resolution of optical focus sensors is limited by the size of the focus spot of the microscope objective, typically about 1 µm. The measured surface height can not be better than the average height of the surface over the spot size [Crea92]. This means that the smallest measureable features are about 2 µm. It has also be taken into account that the sensor works with reflected light. This means that too deep slopes of the surface may not be detected correctly due to the limited numerical aperture of the microscope objective.

Fig. A.3 – Schematics of a profilometer with an optical focus sensor
(After [Crea92])

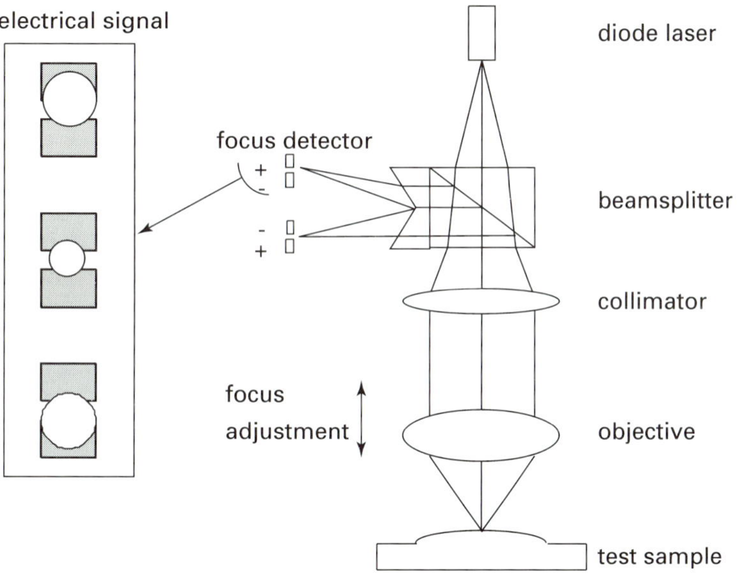

Vertical scanning interferometer

An interference microscope built up as Mirau interferometer is shown in fig. A. 4. Using white light illumination the modulation of the fringes drops rapidly because of the very short coherence length. Only if the opitcal path difference in the two arms is minimum, the modulation signal has a narrow maximum. A measurement of the relative height of a test sample is possible if the modulation peak is detected for every scan position. Fig. A. 5. shows a result for a multilevel Fresnel lens measured with this technique.

Apendix: Testing of Micro-optical Elements

Fig. A.4 – Mirau interference microscope (After [Cabe94])

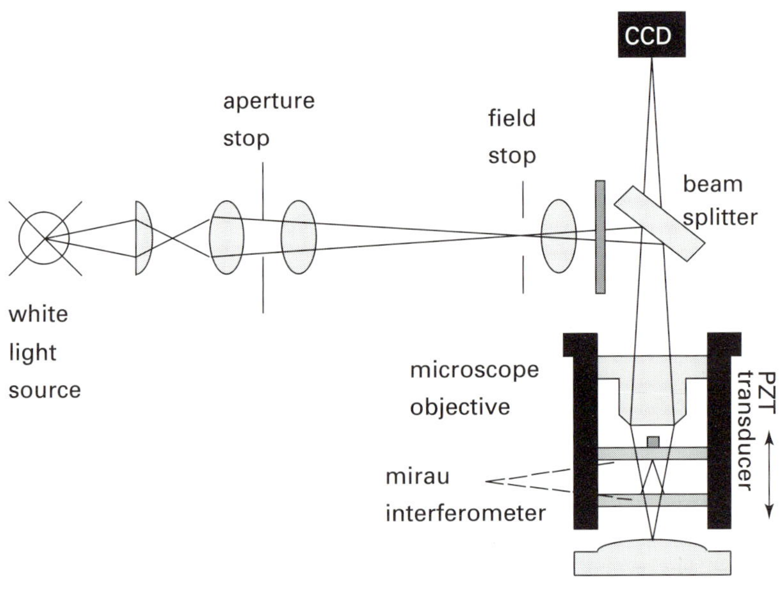

Fig. A.5 – Surface scan of a multilevel Fresnel Zone plate array
(Courtesy Nancy Nieuborg)

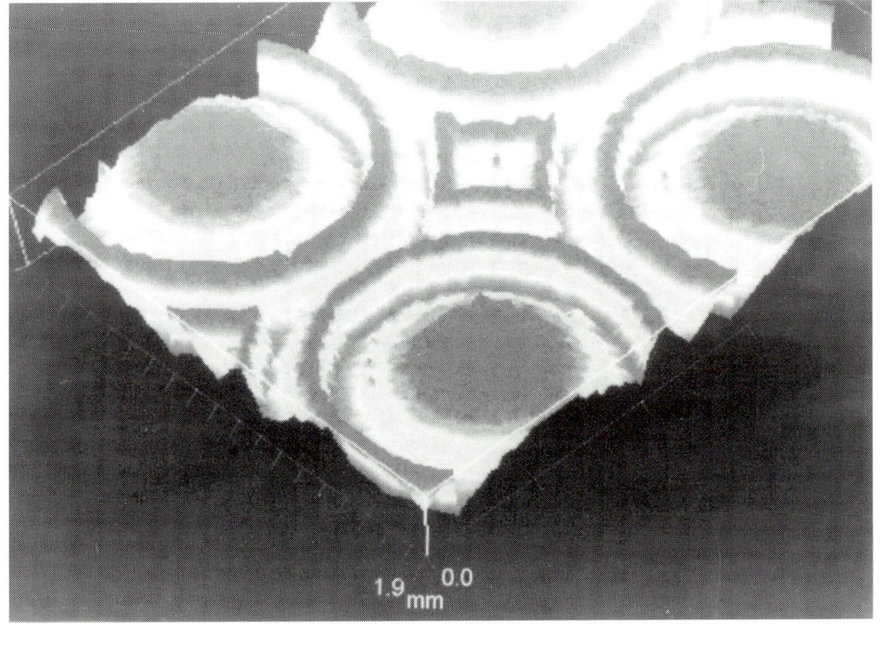

A.1.3. Interferometric testing

Another optical noncontact testing method is based on interferometry. Two basic principles do exist: reflection and transmission interferometers, correlated to the way how the optial element under test lateron will be used. Surface measurement is to be done best in reflection; transmission interferometers are discussed later. In both cases a phase shifting algorithm has to be applied to calculate the phase information from the interference fringes, which basically work as follows [Bren93]:

Principle of phase shifting

The intensity after a two beam interference can be described as

$$I(x,y) = I_0(x,y) [1 + V(x,y)\cos(\Phi(x,y) - \varphi_r)] \tag{12.1}$$

where I_0 is the local average intensity, V is the visibility, φ is the local phase of the sample reflex, and φ_r is the phase of the mirror reflex. The phase distribution of the sample can be computed from four interferograms in phase-quadrature as

$$\phi(x,y) = \arctan(\frac{I_2 - I_4}{I_2 - I_3}) \; (\text{mod } \pi) \tag{12.2}$$

where I_r, $r = 1..4$, are the interferogram intensities corresponding to the reference phases $\phi_r = 0, \pi/2, \pi, 3\pi/2$.

For both cases of reflection and transmission the phase can be evaluated on a computer equiped with a frame grabber. Problems can appear in case of discontinuities of the surface under test with an optical path difference of more than π.

Reflection interferometer - Linnik interference microscope

Fig. A. 6. shows a setup for a Linnik interferometer. It is a measuring tool for relatively flat surfaces. The incoming laser beam is splitted into an object and a reference beam. The light in the object path is reflected from the surface of the test sample (eg. glass, PMMA: 4 % reflection intensity). The reference surface can be a flat mirror (> 90% reflection intensity). Since the reflected light in both arms of the interferometer do not have equal intensity, polarization optics can be used to compensate. Examples of such interferograms have already been shown in fig. 9.3,

10.11 and 10.13. In these examples flat prism surfaces have been measured. Problems always can occur from unwanted back reflection from the back side of the test sample. If surface microlenses should be tested then another type of interferometer may be used. It is called Twyman-Green interferometer with a spherical reference surface. The principle setup is the same as in fig. A.6 [Schw95].

Fig. A.6 – Schematic of a Linnik interferometer (After [Bren93])

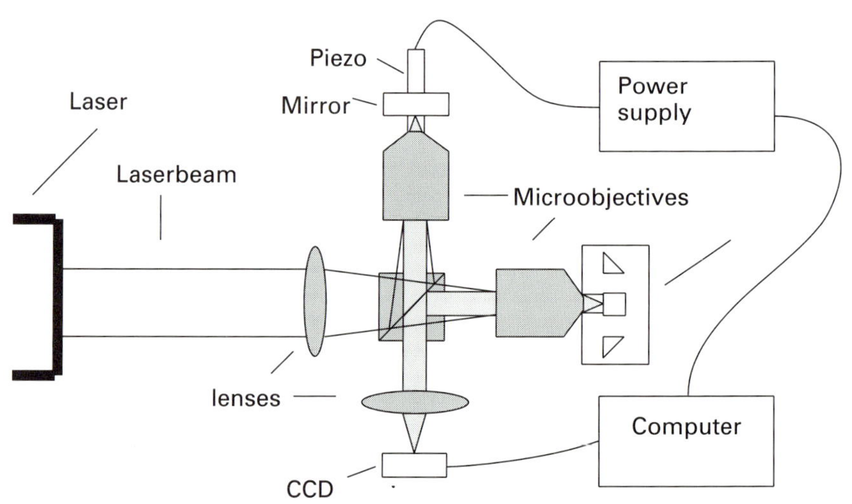

A.2. Measurement of optical functions

A.2.1. Microscope

If the microscope is equiped with micrometer screws for the height, then simple focal length measurement is possible as shown in fig. A. 7. When focussing to the lens surface and afterwards to the image of the light bulb of the microscope formed by the microlens, the difference in height gives the front vertex focal length.

Fig. A.7 – The measurement of front vertex focal length.

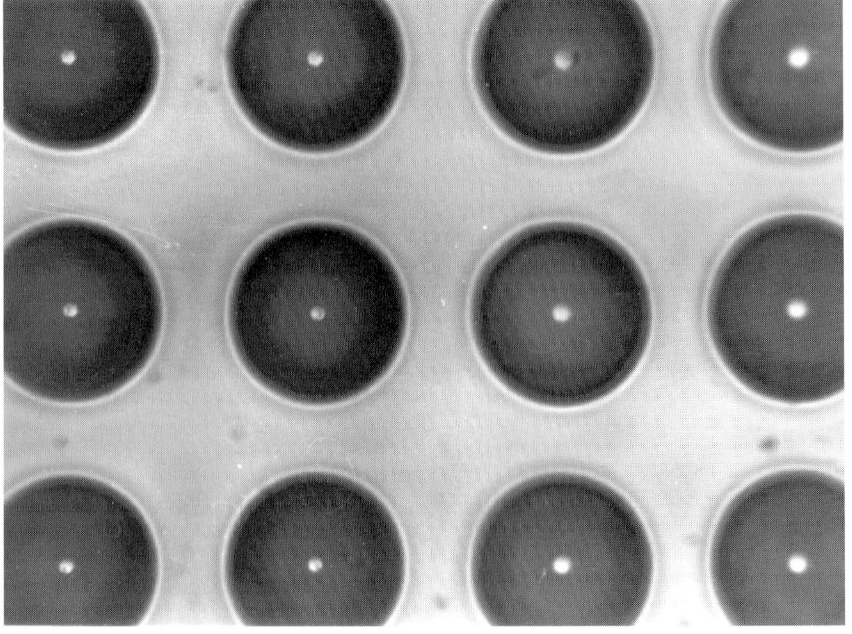

Fig. A.8 – Look to the focal points of GRIN microlenses through a microscope

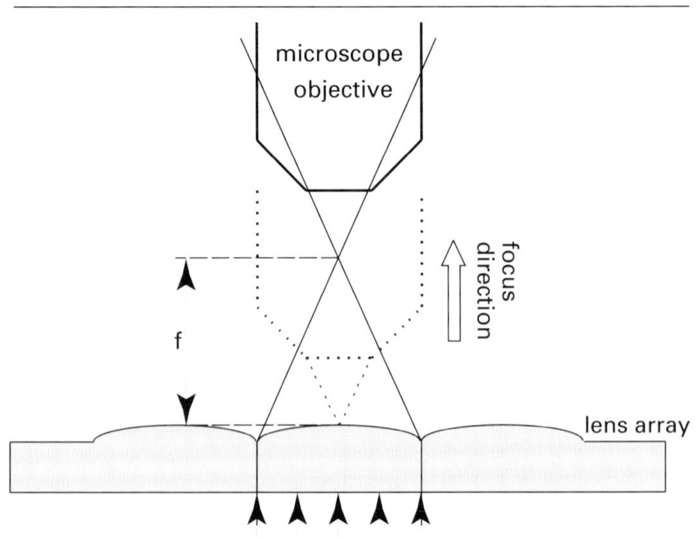

A. 2. 2. Imaging setup

The imaging properties of a microlens can also be tested directly in a setup shown in fig. A. 9. In this setup a microscope objective projects a real demagnified image of the mask pattern in front of the microlens. If the microlens is a distance 2f away from this image then a second real 1:1 image will be produced in a position 2f behind this microlens. The second microscope objective takes this image and transfers it to the CCD camera. This setup can also be used for the determination of the focal length.

Fig. A.9 – Setup for the measurement of imaging characteristics of microlenses (After [Kufn93a])

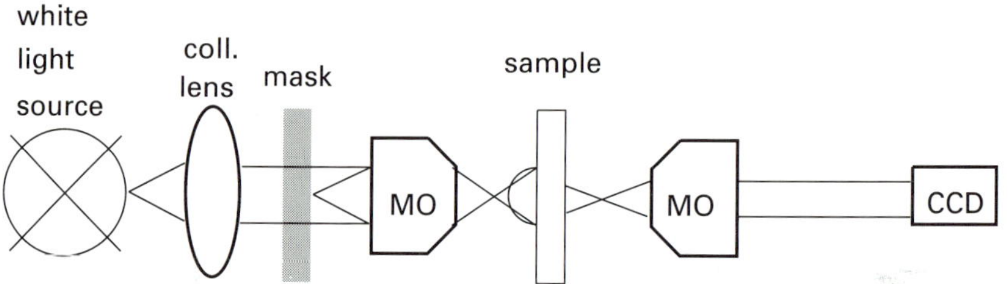

But in addition it is useful to test spatial resolution and image distortion. Fig. A. 10. shows an examples of a typical test pattern. It is a bar patteren with different spatial frequencies.

Fig. A.10 – Image experiment with bar patterns from 8 to 256 lp/mm; left: image formed by the microlens, right: control pattern without microlens. [Kufn93a]

This setup allows also to measure the point spread function of a microlens. The image of an ideal point aperture can never be better than the diffraction limited Airy disk. The result of such an experiment is shown in fig. A. 11. The intensity in this image can be described by the following formula:

$$I = (\frac{J_1(r)}{r})^2, \quad r = \frac{D\pi}{2f\lambda} \tag{12.3}$$

J_1 is the Besselfunction of 1stOrder, D is the diameter of the microlens, f is the focal length and λ is the wavelength. This corresponds to an aperture with the same diameter than the lens. The zone inside the first zero corresponds to 84% of the total intensity. The quotient between real and ideal intensity is called Strehl ratio. A Strehl ratio better than 0.3 is in most cases a useful microlens.

Another characteristic function is the modulation transfer function (MTF). It describes the relation between spatial frequency and contrast. The contrast (or the modulation) M can be defined as

$$M = \frac{I_{max} - I_{min}}{I_{max} + I_{min}} \tag{12.4}$$

Fig. A.11 – Focus of microlens with a Strehl ratio of about 0.5 [Kufn93b]

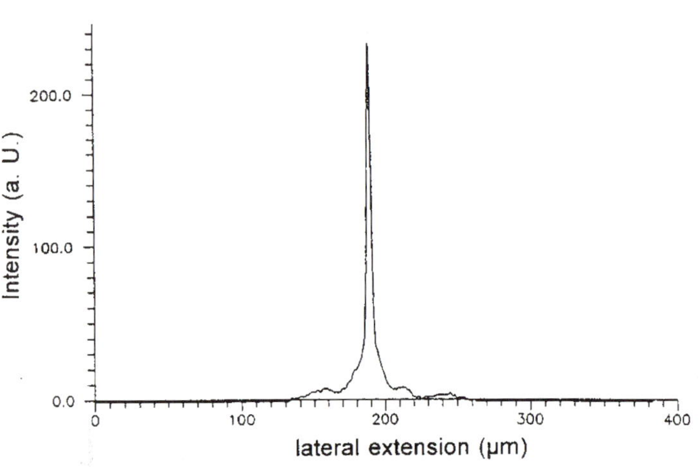

Fig. A. 12 illustrates the meaning of the MTF. It shows a bar pattern with increasing spatial frequency in one dimension (the transparent parts of the mask are decreasing in width from left to the right). The ideal image should look like fig. A.11a). But in reality the contrast is decreasing with increasing spatial frequency as shown in fig. A.11b). In fig. A.16 a MTF is shown. The dashed line corresponds to the diffraction limit.

Fig. A.12 – Illustration of the decreasing contrast with increasing spatial frequency

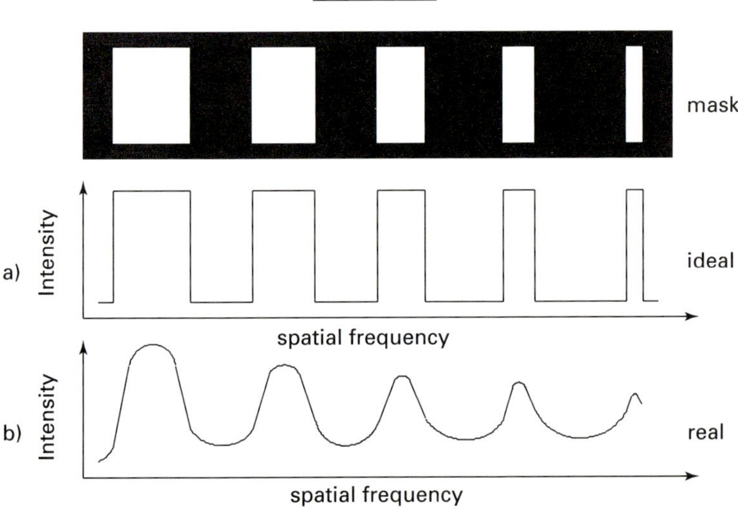

Fig. A.13 – Schematic of a Mach-Zender interference microscope

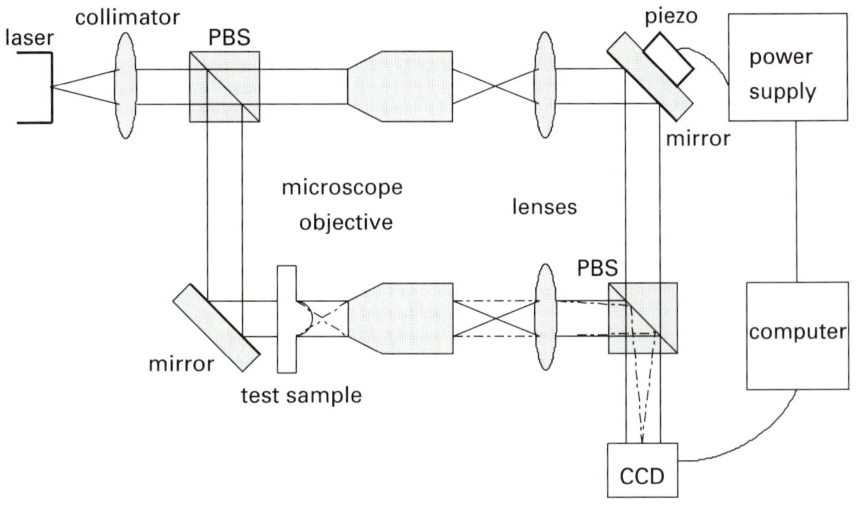

A. 2. 3. Transmission interferometer

Fig. A. 13. shows a setup for a Mach-Zender interferometer. It is especialy usefull for the measurement of microlenses because the complete transmission information (surface influence and index of refraction) is obtained. Therefore it can be used for surface microlenses as well as for gradient index microlenses. As an example fig. A.14 shows an interferogram of surface microlenses.

Fig. A.14 – Interferogram of surface microlenses
with a diameter of 200 µm.

In the interferometer the incoming laser beam is splitted into an object and a reference beam. The two seperated paths are brought to interference and are recorded on the CCD camera. With the phase shifting technique as described above the relative phase information can be calculated. Fig. A. 15 shows the optical path difference (OPD) for a refractive microlens profile. With the appropriate software it is also possible to determine the modulation transfer function as mentioned earlier. Fig. A.16 shows a typical plot.

Micro-optics and Lithography

Fig. A.15 – Calculated OPD of a refractive microlens [Kufn93b]

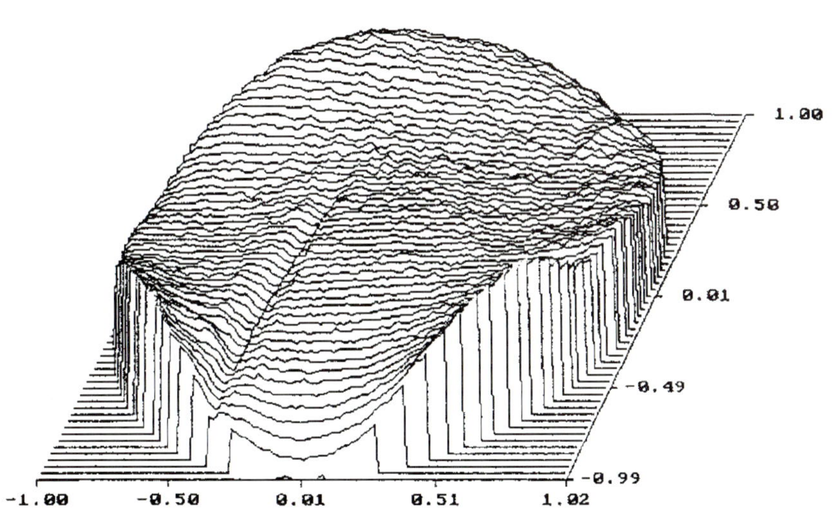

Fig. A.16 – MTF of the microlens from fig. A.14 [Kufn93b]

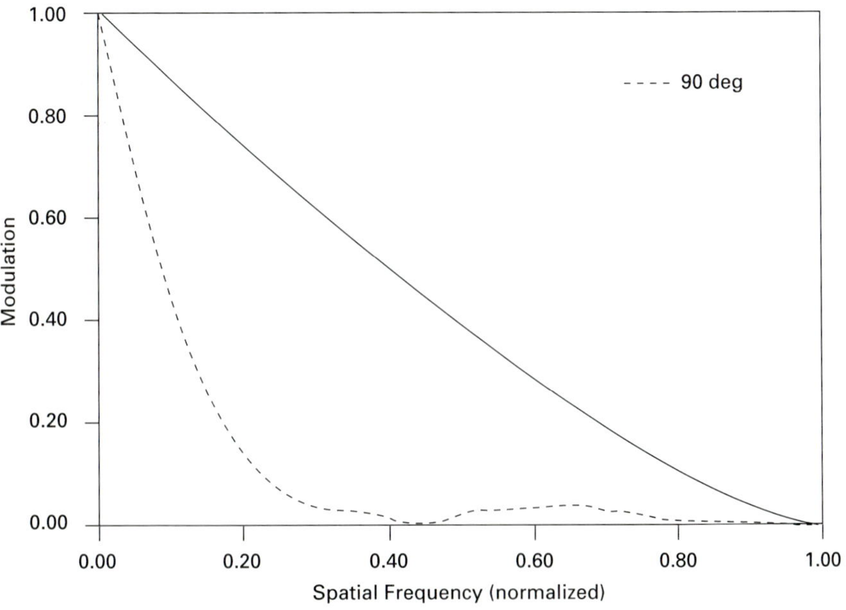

References for Appendix A

[Bren93]   K.-H. Brenner, M. Kufner, S. Kufner, J. Moisel, A. Müller, S. Sinzinger, M. Testorf, J. Göttert, and J. Mohr: "Application of three-dimensional micro-optical components formed by lithography, electroforming, and plastic molding", Appl. Optics 32 (32), 6464-6469, 1993.

[Cabe94]   P. J. Caber, S. J. Martinek, and R. J. Niemann: 3A new interferometric profiler for smooth and rough surfaces", in WYKO Product information, Tucson, 1994.

[Crea92]   K. Creath and A. Morales: "Contact and noncontact Pofilers", in 'Optical Shop Testing', chapter 17, ed. D. Malacara, Wiley, New York, 1992.

[Kufn93a]   M. Kufner, S. Kufner, M. Frank, J. Moisel, M. Testorf: "Microlenses in PMMA with high relative aperture: A parameter study", Pure Appl. Opt. 2 (1993) 9-19.

[Kufn93b]   M. Kufner: "Herstellung und Charakterisierung von hochgeöffneten Mikrolinsen in PMMA", Thesis Uni Erlangen-Nürnberg, 1993.

[Laza96]   S. Lazare, J. Lopez, J.-M. Turlet, M. Kufner, S. Kufner, P. Chavel: "Microlenses fabricated by UV excimer laser irradiation of PMMA followed by styrene diffusion", Applied Optics, Vol. 35, No. 22, p. 4471 - 4475, 1996.

[Schw95]   J. Schwider and O. Falkenstörfer: "Twyman-Green interferometer for testing microspheres", Optical Engineering, Vol. 34, No. 10, p. 2972 - 2975, 1995.

Keywords

irradiation
- deep proton, 153, 159, 171, 175-176, 185
- synchrotron, 29, 137, 139, 143

L

laser
- beam writing, 44
- scanner, 9, 27, 104, 125, 131, 133, 149, 173, 184
lateral straggling, 156, 163, 173
lift-off, 33
LIGA, 74, 137-151, 154, 178, 184
light path folding, 107
$LiNbO_3$, 54, 61, 67-68, 71, 75-76, 79, 91, 178
lithography, 1, 3, 9-12, 14, 16, 18, 20, 22-40, 42, 44, 46, 48, 50, 52, 54, 56, 58-59, 61-62, 64, 66, 68, 70, 72, 74, 76, 78, 80, 82, 84, 86, 88, 92, 94, 96, 98, 100, 102, 104, 106, 108, 110, 112, 114, 116, 118, 120, 122, 124-126, 128, 130, 132, 134, 136-178, 180-182, 184, 186, 188, 190, 192, 194, 196, 198, 200-201

M

main chain scission, 29, 131
mask
- chromium, 26-27, 88
- gold, 137-139
- grey scale, 45, 133
- plotter, 26, 45
mastering, 44, 69, 105, 137
mechanical mount, 54, 88, 166-168
melting, 11, 39, 73, 91, 97, 113, 115-119, 121, 123, 130-131, 143-144, 151, 178
microbeamsplitter, 140
microlens
- astigmatic, 97-98

- cylindrical, 142
- diffractive, 36, 49, 51, 59, 121
- dispersive, 51-52, 58
- radial GRIN, 74, 98, 100, 110, 194
- refractive, 17, 39, 51, 107, 118, 123, 135, 199-200
micromachining, 58
microprism, 108-109, 171-172
miniaturization, 10-11, 13, 18-19, 21, 23, 105
minimum feature size, 18, 26-27, 42, 44, 46, 48, 84
mode, 61, 66-67, 70-71, 75-76, 87, 148, 173
modulation transfer function, 196, 199
moulding, 69, 80, 137-139, 151

N

numerical aperture, 44, 61, 63, 94-95, 99, 112, 189

O

optical data transcription, 109
optical focus sensor, 189-190

P

packaging, 54, 79
perfect shuffle, 120, 169-170
phase level, 42, 49
phase shifting, 192, 199
photoablation, 125-127, 130-131, 135
photosensitive glass, 81, 83, 85-87, 89, 183
planar optics, 54-57, 102-104, 108, 178-179, 181
PMMA, 28-29, 31, 46, 69-70, 105, 119, 127, 131-135, 137-139, 143,